1st Portuguese Condensed Matter Physics National Conference (CMPNC 2019)

EPJ Web of Conferences Volume 233 (2020)

Porto, Portugal
8 - 10 May 2019

Editors:

Bernardo Goncalves Almeida
J. Agostinho Moreira

ISBN: 978-1-7138-1016-2

Printed from e-media with permission by:

Curran Associates, Inc.
57 Morehouse Lane
Red Hook, NY 12571

Some format issues inherent in the e-media version may also appear in this print version.

For additional information, please contact EDP Sciences – Web of Conferences
at the address below.

EDP Sciences – Web of Conferences
17, Avenue du Hoggar
Parc d'Activité de Courtabœuf
BP 112
F-91944 Les Ulis Cedex A
France

Phone: +33 (0) 1 69 18 75 75
Fax: +33 (0) 1 69 28 84 91

contact-edps@webofconferences.org

Additional copies of this publication are available from:

Curran Associates, Inc.
57 Morehouse Lane
Red Hook, NY 12571 USA
Phone: 845-758-0400
Fax: 845-758-2633
Email: curran@proceedings.com
Web: www.proceedings.com

TABLE OF CONTENTS

Guest Editorial

The 1st Portuguese Condensed Matter Physics Meeting (PCMPM) was held from 8th to 10th may 2019, in the Faculty of Sciences of the University of Porto, under the sponsoring of the Portuguese Physical Society (SPF). The Organizing Committee involved the members of the Condensed Matter Physics Division of SPF and the Program Committee included members of the Universities of Porto, Minho, Aveiro, Coimbra, Lisboa and Nova de Lisboa, where intense and internationally recognized research on the area is being carrying out.

The central aim behind of this conference was to bring together young and senior researchers in Condensed Matter Physics, in order to share experiences, to spread the latest information on progress in their specialties and related fields, to gain visibility for their research, and to allow young scientists to develop professionally. The meeting aims at constructing a broad space, towards a profound discussion and reflection regarding the past, present and future of this research field in Portugal, and an assessment of its international impact.

One hundred and fifteen attendees from different Portuguese Universities and Polytechnics were fully registered at PCMPM, making this meeting a truly Portuguese initiative. These participants took active part in the sessions and informal discussions that had place along the three days of the meeting. There were eight plenary talks, given by Eduardo V Castro, Michael Belsley, L. Pereira, Rui Travasso, Nuno A M Araújo, Ricardo Ferreira, Susana Cardoso, and João Ventura, forty four oral and forty six poster contributions, covering a wide number of topics ranging from fundamental research to new characterization techniques and device physics, mirroring all research areas in Portugal. It is worth to stress that sixty five MSc and PhD students were registered in this conference.

Complementary to the program of the meeting, a one day satellite "Young Mind Project" Meeting took place on Thursday 9th, fully dedicated to undergraduate physics students coming from all Portuguese Universities. This meeting was introduced by Dr. Roberta Caruso from Naples University and Dr. Teresa Coimbra, from University of Aveiro. They presented the Young Mind initiative to the Physics Students' Nuclei, in order to spread and strength the project in Portugal, enabling a broader participation of our students in the European Physical Society (EPS) activities.

PCMPM is the first of a series of meetings that we aim at organizing every two years. The 2-year cycle makes them very attractive and enables the opportunity to update the actual knowledge, through intensive scientific programs with lectures and contributions delivered by well-known speakers, experienced experimentalists and theoreticians. We believe that the PCMPM will be an excellent opportunity to report innovative ideas and results, allowing interesting and fruitful discussions, and explore new trends in the extended research field of condensed matter physics.

We would like to express our warmest thanks to all the participants who contributed to this publication, as well as to the referees that helped selecting and improving the manuscripts. Our thanks are extended to the Scientific Committee, to the Directory Board of SPF and to the Dean of the Faculty of Sciences of the University of Porto. A special thanks to Fundação Engenheiro António de Almeida, for the financial support of this publication. We hope that these proceedings may be useful not only to researchers directly engaged in the field, but also to the larger community working in condensed matter physics and material science. Finally, we express our gratitude to all the public institutions and companies that, in many ways, contributed to the success of PCMPM.

B. Almeida

J Agostinho Moreira

Statement of Peer review

In submitting conference proceedings to EPJ Web of Conferences, I certify to the Publisher that I adhere to the **Policy on Publishing Integrity** of the journal in order to safeguard good scientific practice in publishing.

1. All articles have been subjected to peer review administered by the proceedings editors.
2. Reviews have been conducted by expert referees, who have been requested to provide unbiased and constructive comments aimed, whenever possible, at improving the work.
3. Proceedings editors have taken all reasonable steps to ensure the quality of the materials they publish and their decision to accept or reject a paper for publication has been based only on the merits of the work and the relevance to the journal.

Title, date and place of the conference

Portuguese Condensed Matter Physics Meeting, Porto, 8 – 10 May, 2019. Faculdade de Ciências, Universidade do Porto.

Proceedings editor(s):

Bernardo Gonçalves Almeida e J Agostinho Moreira

Date and editor's signature

Porto, 29th March 2020

Quantum effects in electrical transport properties of Bismuth chalcogenides Topological Insulators

José A. Paixão[1,*], *Marta S.C.* Henriques[1], *Carlotta* Micale[2], *Elsa B.* Lopes[2], *Vanda M.* Pereira[3], and *António P.* Gonçalves[2]

[1]CFisUC, Department of Physics, University of Coimbra, P-3004-516 Coimbra, Portugal
[2]C²TN, DECN, Instituto Superior Técnico, University of Lisbon, Estrada Nacional 10, P-2695-066 Bobadela LRS, Portugal
[3]Max-Planck Institute for Chemical Physics of Solids, Nöthnitzer Str. 40, D-01187 Dresden, Germany

Abstract. Quantum effects such as weak-antilocalisation (WAL) behavior and Shubnikov–de Haas (SdH) oscillations in the electrical transport properties of topological insulators, measured on nanostructured polycrystalline samples and single-crystals of a series of bismuth chalcogenide compounds ($Bi_2(Se_xTe_{1-x})_3$, $0 \leq x \leq 1$ and $BiSbTe_3$), are presented and discussed.

1 Introduction

Topological insulators (TIs) are celebrated novel materials with a peculiar band structure that features a gap for bulk states but may also host spin-polarised surface conducting states, topologically protected from scattering mechanisms that do not break time-inversion symmetry [1]. A representative family of such materials are bismuth chalcogenides, that have attracted a lot of interest also for their good properties as thermoelectric compounds [2].

The absence of back-scattering of the 2D electrons by non-magnetic impurities on TIs gives rise to a distinctive phenomenon known as weak anti-localisation (WAL) effect, a quantum correction to the classical magnetoresistance. It is characterised as a sharp cusp of the 2D magnetoconductance $\Delta\sigma_{xx}(B)$ at zero magnetic field. This quantum effect is well described in 2D systems by the Hikami-Larkin-Nagaoka (HLN) equation [3],

$$\Delta\sigma_{xx} = \alpha\frac{e^2}{\pi h}\left[\Psi\left(\frac{\hbar c}{4el_\phi^2 B} + \frac{1}{2}\right) - \ln\left(\frac{\hbar c}{4el_\phi^2 B}\right)\right], \tag{1}$$

where Ψ is the digamma function, l_ϕ is the phase coherence length, and the prefactor α is a coefficient with a negative value close to $-1/2$ for the WAL, arising from the gapless surface states of TIs.

At high magnetic fields and low temperatures, SdH oscillations are often found in the electrical transport properties of TIs, as the Fermi level crosses the Landau levels of the 2D electron gas. Analysis of such oscillations provides invaluable information about the Fermi surface. The Lifshitz-Onsager equation relates the index n of the SdH oscillations to the cross-section area of the Fermi surface perpendicular to the applied field,

$$2\pi(n + \gamma) = \pi k_F^2 \frac{\hbar}{eB}, \tag{2}$$

*Corresponding author: jap@fis.uc.pt

where πk_F^2 is the cross-sectional area of the Fermi surface perpendicular to B with radius equivalent to the Fermi wave vector k_F, and $\gamma = 1/2 - \beta - \delta$ is the phase of the SdH oscillations (in 2π units) of the σ_{xx} component of the electrical conductivity tensor, with β the Berry phase and δ a phase shift correction determined by the dimensionality of the system. For conventional metals, $\beta = 0$, and for Dirac systems it should be $\frac{1}{2}$. For a 3D electron system $\delta = \pm 1/8$, while for a 2D system $\delta = 0$ [4]. Experimentally, the Berry phase can be obtained by plotting the index n of the oscillations (position of the σ_{xx} minima) versus $\frac{1}{B}$ [5]. The amplitude of the SdH oscillations decreases rapidly with temperature following the Lifshitz-Kosevich law [6]. By measuring and analysing the temperature dependence, an important piece of information, the effective mass of the charge carriers can be obtained.

2 Materials and Methods

Nanostructured samples of $Bi_2(Se_xTe_{1-x})_3$, $0 \leq x \leq 1$, were synthesised using a simple wet-chemical route previously reported [5, 7]. Stoichiometric amounts of bismuth nitrate pentahydrated, sodium selenide and/or telluride and potassium hydroxide were dissolved in ethylene glycol. The solution was heated in a microwave digestion system with power, target temperature, heating and cooling times adjusted to achieve the best results. The reaction product is a precipitate that was centrifuged, washed with deionised water and ethanol, and later dried. Purity of the samples was checked by powder X-ray diffraction on a Bruker D8-Advance diffractometer (Fig. 1).

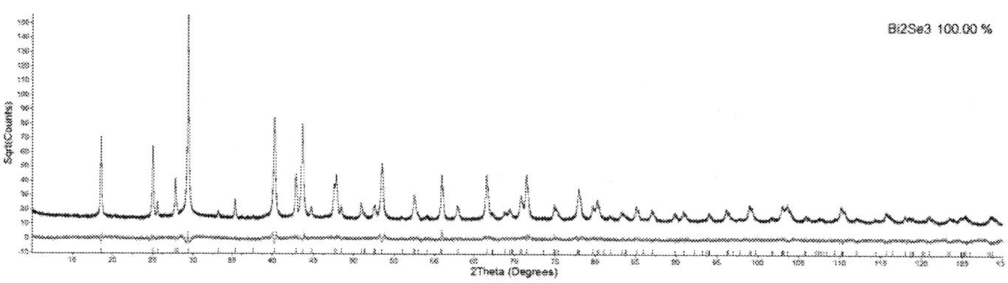

Figure 1. Powder X-Ray diffractogram of a nanostructured Bi_2Se_3 sample. No extraneous Bragg peaks of impurities were detected (the y axis scale is depicted as the square-root of the intensity for better sensitivity to small peaks). The solid line passing through the measured points is the result of a Rietveld fit, the flat difference line between the measured and calculated intensities shown at the bottom further proves the sample purity.

Although this is a simple and fast synthetic method, fine adjustment of the heating conditions is necessary because extraneous phases are easily formed e.g. Bi and Te oxides, Bi_2Se_2Te and unreacted Te. Under optimised conditions, easier to attain for the end-member compositions, single-phase samples can be produced in the form of small nanoplates of hexagonal habit of micrometer size and nanometer width (Fig. 2).

Single crystals of Bi_2Te_3 and $BiSbTe_3$ were grown by the Bridgman-Stockbarger method, starting from high purity elements (6N) that were further purified by floating zone (Bi,Sb) and sublimation (Te) refinement before attempting crystal growth. As the electronic properties of these compounds are critically dependent not only on even small amounts of impurities but also on stoichiometry, slightly different deviations from the ideal composition were tried. Small pieces of the obtained crystals were tested to confirm the correct structure and composition using a single crystal X-Ray diffractometer (Bruker ApexII) and a SEM/EDS (Vega3 SBH, Tescan) equipment.

Figure 2. SEM image of the Bi_2Se_3 crystals produced using a wet-chemical method (see Sect. 2.)

3 WAL in nanostructured Bismuth chalcogenides

A strong WAL effect is a distinctive quantum phenomenon typically found in bismuth chalcogenide TIs. It has been observed in single-crystalline, polycrystalline and thin film samples. The electrical transport properties of our nanostructured samples were measured on small pellets formed by pressing the dried synthesised powder under a 6 MPa pressure. XRD data show that these pellets are strongly textured with the hexagonal *c*-axis perpendicular to the pellet surface. The electrical conductivity and Hall effect of such samples were measured as function of temperature ($1.8 \leq T \leq 300$ K) and magnetic field ($0 \leq B \leq 9$ T) on a Dynacool Quantum Design PPMS system. A particular attention needed to be taken to the electrical sample contacts, as the usual procedure of directly gluing thin gold wires into the sample with conducting silver paint resulted in a time-dependent resistivity arising from thermal diffusion of Ag atoms into the sample. Therefore, the electrical connections to the sample were made on Au/Pd sputtered contacts on the sample surface that proved effective to strongly reduce the above mentioned diffusion effect. Fig. 3 shows the measured magnetoresistance at 1.8 K of the pure nanostructured Bi_2Se_3 sample discussed in Sect. 2.

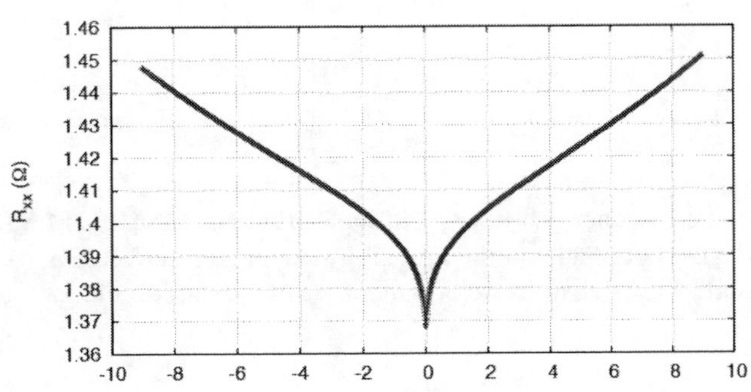

Figure 3. Magnetoresistance of a nanostructured Bi_2Se_3 sample measured at 1.8 K, showing a strong weak-antilocalisation effect.

https://doi.org/10.1051/epjconf/202023301001

Interestingly, WAL is a robust phenomenon that can still be easily observed in similar nanostructured samples even if they contain a significant amount of extraneous phases [5]. This surprising observation can be explained by the large coherence length of the electronic wave-functions at low temperatures. Its temperature dependence (Fig. 4) provides additional information regarding the dimensionality of the electron system: for 2D systems, and a dominant electron-electron dephasing mechanism, it is expected a power-law dependence $l_\phi \sim T^{-1/2}$, while for 3D systems $l_\phi \sim T^{-3/4}$ [9].

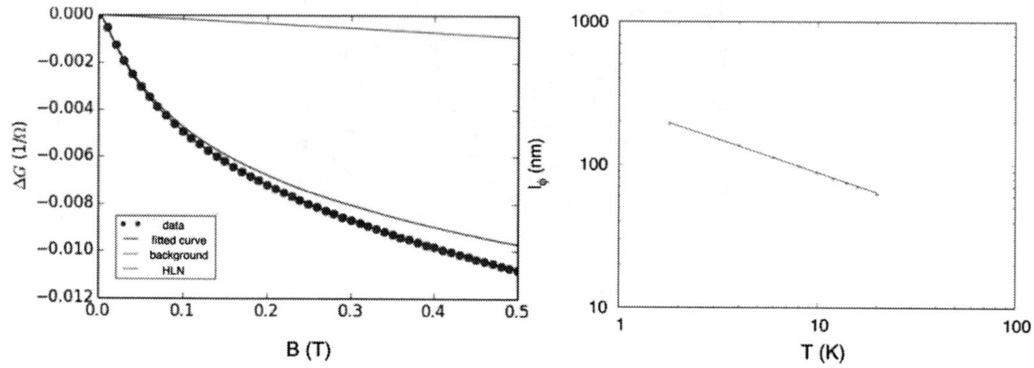

Figure 4. (left) Fit of the HLN model to the $\Delta G = 1/R_{xx}(B) - 1/R_{xx}(0)$ data of the nanostructured sample of Bi_2Se_3 measured at 1.8 K. The fit yields a coherence length $l_\phi = 193.96 \pm 0.17$ nm at 1.8 K. (right) Power-law fit of coherence length dependence on temperature. The fit yields $l_\phi \propto T^{-0.463(5)}$, corroborating that the WAL effect originates from 2D electronic states. The fitted $\alpha = -3.67(1)$ value deviates significantly from the expected vale of -1.0 for two 2D conducting channels, at the top and bottom surfaces, but this parameter is prone to error due to the large uncertainty in estimating the geometrical factors when converting the measured bulk resistance into the 2D sheet conductance.

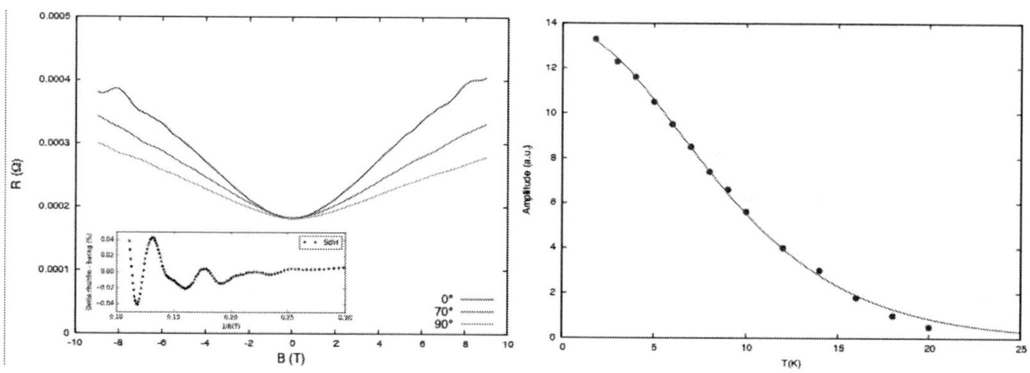

Figure 5. (left) Magnetoresistance measured on a single crystal of $BiSbTe_3$ at 1.8 K for different angles between the applied magnetic field and the normal to the sample surface, showing SdH oscillations. (right) Fit of the Lifshitz-Kosevich model to the amplitude of the SdH oscillations, giving $m^\star = 0.14\, m_0$.

4 Shubnikov-de Haas oscillations

SdH oscillations in the electrical transport properties have been often observed in bismuth chalcogenide compounds. In contrast with the WAL behaviour, SdH oscillations are easier to observe when the samples depart from the typical semiconductor behaviour into the metallic regime, due the position of the Fermi level at a specific stoichiometry or to the presence of impurities that may act as electron donors or acceptors. We illustrate this phenomenon on data from a single crystalline $BiSbTe_3$ sample grown as described in Sect. 2. The electrical resistivity shows a typical metallic behaviour with $\rho \sim 16\ \mu\Omega\,\mathrm{m}$ and a $RRR \sim 15$. The measured thermoelectric power at room temperature is $S = 130\ \mu\mathrm{V/K}$. The positive sign of the thermopower as well as the sign of the Hall coefficient show that the majority electric carriers are positively charged. i.e., are holes in the valence band.

The magnetoresistance measured as function of the applied field did not exhibit the low-field cusp associated with WAL but at high fields SdH oscillations of the magnetoresistance are clearly observed. These oscillations are, as expected, stronger when the field is perpendicular to the sample surface and tend to vanish if it is parallel to the surface. The amplitude of the SdH oscillations was followed as function of temperature. The data shows a good fit to the Lifshitz-Kosevich law, as depicted in Fig. 5. The cyclotron effective mass of the dominant charge carriers obtained from such fit is $m^\star = 0.14\ m_0$, close to the calculated value of the hole mass in the valence band of $BiSbTe_3$ [10].

5 Conclusions

We have observed distinctive quantum effects (WAL and SdH oscillations) in the electrical transport properties of TIs of the bismuth chalcogenides family. WAL is easier to find in those samples that are less conducting, and could be observed in nanostructured samples as well as in single-crystals. It is a particularly robust effect that was observed both in pure and impure samples. For the most conducting samples, the SdH effect was clearly observed and was analysed within the framework of the Lifshitz-Onsager model. The temperature dependence of the amplitude of the SdH oscillations follows the Lifshitz-Kosevich law [6]. Other quantum effects such as the Aharonov-Bohm effect have been recently reported [8] in $BiSbTe_3$ adding to the interest of studying such TI compounds.

This work was partially supported by funds from FEDER (Programa Operacional Factores de Competitividade COMPETE) and from FCT under the projects UID/FIS/04564/2016 and PTDC/FIS-NAN/6099/2014. Access to instruments of the TAIL-UC facility funded under QREN-Mais Centro project ICT_2009_02_012_1890 is also gratefully acknowledged.

References

[1] Y. Hondo, J. Phys. Soc. Jpn. **82**, 102001 (2013).
[2] D. Park, S. Park, K. Jeong, H-S. Jeong, J.Y. Song, M-H Cho, Scientific Reports **6**, 19132 (2016).
[3] S. Hikami, A. I. Larkin and Y. Nagaoka, Prog. Theor. Phys. **63**, 707 (1980).
[4] D. Kumar, A. Lakhani, Phys. Status Solidi RRL **9(11)** 636–640 (2015).
[5] V.M.M. Pereira, M.S.C. Henriques, J.A. Paixão, Physica B **536** 51-55 (2018).
[6] I.M. Lifshitz, A.M. Kosevich, Sov. Phys. JETP **2**, 636 (1955).
[7] H. Xu, G. Chen, R. Jin, D. Chen, Y. Wang, J. Pei, Y. Zhang, C. Yan, Z. Qiu, CrystEngComm **16**, 3965-3970 (2014).
[8] S. Huang, P. Wang, C. Lin, S. You, W. Lin, L. Lin, Y. Yan, S. Yu, M. Chou, Appl. Phys. Lett. **112**, 203103 (2018).
[9] J. Liao, Y.B. Ou, H.W. Liu, K. He, X.C. Ma, Q.K. Xue, Y.Q. Li, Nat. Commun. **8**, 16071 (2017).
[10] B.S. Farag, A.M.A. El Soud, H.A. Zayed, S.A. Gad, J. Ovonic Res. **6**,267-275 (2010).

Magnetic studies of monoclinic $Cu_4O(SeO_3)_3$, a copper-oxo-selenite derivative

José F. Malta[1,2,*], *Marta S. C.* Henriques[1], *José A.* Paixão[1], and *António P.* Gonçalves[2]

[1]CFisUC, Department of Physics, University of Coimbra, Rua Larga, 3004-516 Coimbra, Portugal

[2]C²TN, DECN, Instituto Superior Técnico, University of Lisbon, Estrada Nacional 10, 2695-066 Bobadela LRS, Portugal

Abstract. $Cu_4O(SeO_3)_3$ is a copper-oxo-selenite belonging to the $Cu_xO(SeO_3)_{(x-1)}$ family of the topological chiral magnet Cu_2OSeO_3. We report magnetometry and specific heat data measured in a monoclinic $Cu_4O(SeO_3)_3$ single crystal grown through a Chemical Vapour Transport (CVT) reaction. Our study shows a typical antiferromagnetic behaviour, with a Néel temperature T_N = 58 K, similar to that of the Cu_2OSeO_3 and an additional transition at 13 K. The effective magnetic moment per Cu atom is 1.84 μ_B, close to the expected theoretical value for Cu^{2+}. The low-temperature $M(H)$ curves, show a transition starting at H_{c1} ~ 400 Oe at 1.8 K shifting to a lower value of ~ 280 Oe at 30 K, likely from a helical into a conical intermediate phase, and a second transition at H_{c2} ~ 1 kOe, above which the net moment increases linearly with applied field. The magnetisation moment value in a 90 kOe field is 0.053 μ_B/Cu at 1.8 K and attains a maximum value of 0.061 μ_B at 13 K. Low-temperature specific heat measurements confirm the presence of the magnetic transition at 13 K, slightly shifting to lower temperatures under an applied magnetic field.

1 Introduction

$Cu_4O(SeO_3)_3$ is copper-oxo-selenite derivative belonging to the family of $Cu_xO(SeO_3)_{(x-1)}$, $4 \leq x \leq 2$ compounds that includes the topological chiral magnet Cu_2OSeO_3, a most celebrated compound where a skyrmionic magnetic phase was found [1]. Cu_2OSeO_3 crystallises in the chiral space group $P2_13$, the same as B20 compounds where skyrmions where first observed [2]. A magnetic skyrmion is a swirling magnetic spin texture carrying a topological quantum number that occurs due to the competition between isotropic Heisenberg ferromagnetic exchange and antiferromagnetic Dzyaloshinskii-Moriya (DM) interactions [3]. In the magnetic ground state of Cu_2OSeO_3, the Cu moments order in a long period (~ 65 nm) helical structure (T_N = 58 K), but close to T_N and under a small applied magnetic field a new magnetic phase with a skyrmionic magnetic arrangement of the spins was found [1, 4].

$Cu_4O(SeO_3)_3$ is known to crystallise in two different polymorphs, with monoclinic and triclinic structures reported long ago by Effenberger et al. [5], the monoclinic structure being more complex than the triclinic one. Although the selenite anions have the typical tetrahedral geometry, the Cu atoms feature distinct coordination geometries. The compound can

*Corresponding author: josemalta@ctn.tecnico.ulisboa.pt

be obtained by thermal decomposition of $CuSeO_3$ in air at high temperatures [6] or using Cu_2OSeO_3. It has also been found as a contaminant during the synthesis of Cu_2OSeO_3 [7, 8]. In this work, we obtained high-quality monoclinic $Cu_4O(SeO_3)_3$ single crystals when we tried to grow Cu_2OSeO_3 using a CVT reaction. Despite the crystal structures of the two polymorphs of $Cu_4O(SeO_3)_3$ having been reported in the literature, their physical properties are poorly known. In particular, magnetic studies deserve to be performed to better understand the nature of the magnetic interactions in a compound closely related to the topological chiral magnet Cu_2OSeO_3.

2 Materials and Methods

Cu_2OSeO_3 powder was obtained by reacting CuO and SeO_2 (2:1 molar proportion) in evacuated sealed quartz tubes heated at 600 °C during 3 days. To avoid the formation of other phases, the tube was quenched in water three times during the synthesis, once each 24 h. An olive-green powder was obtained in the quartz tube that was analysed using powder XRD. The diffractogram confirmed the presence of Cu_2OSeO_3 as a majority phase (90% w/w), unreacted CuO (10% w/w) and no further extraneous phases. This powder was used for the CVT reaction where single crystals were grown.

The previously obtained powder was sealed in evacuated quartz tubes with an additional small portion of NH_4Cl (0.5 g/cm^3) to act as transport agent in the CVT reaction. The tube was placed in a dual zone furnace and high quality crystals (sizes 1-3 mm) were obtained in the colder zone after six weeks of growth. The temperatures of source and deposition zones were 640 °C and 530 °C, respectively.

3 Structural characterisation

A small crystal was selected for data-collection in a single crystal X-ray diffractometer (Bruker APEXII) equipped with a CCD detector and a Mo X-ray tube (λ = 0.71073 Å). The crystal was found to be a specimen of the monoclinic polymorph, with cell parameters a = 15.990(10), b = 13.518(8), c = 17.745(12) Å, β = 90.49(5)° and space group $P2_1/a$. A powder sample was also examined by XRD to check for the presence of extraneous phases. The observed Bragg peaks are consistent with the sole presence of the monoclinic $Cu_4O(SeO_3)_3$ phase.

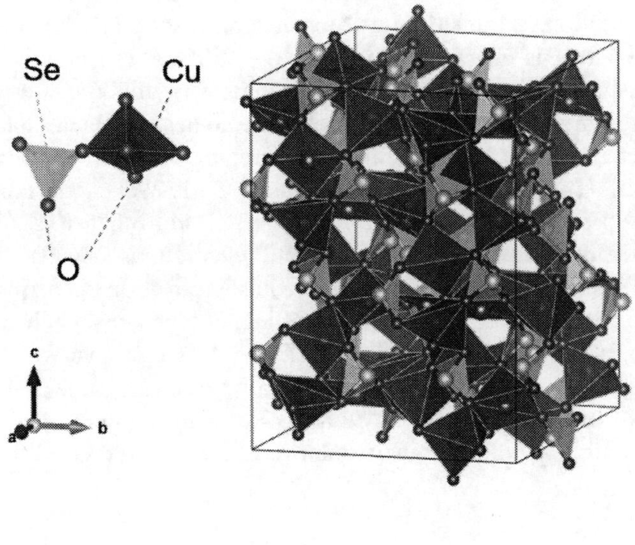

Figure 1. Structure of monoclinic $Cu_4O(SeO_3)_3$. Cu and Se atoms and their polyhedra are coloured blue and green, respectively. The O atoms, located at the corners of the coordination polyhedra, are coloured red.

The monoclinic $Cu_4O(SeO_3)_3$ structure is rather complex. The unit cell depicting the main structural units, Se and Cu coordination polyhedra, is shown in Fig. 1 . The asymmetric unit contains 68 independent atoms. The selenite anions have the typical geometry, and the Cu atoms coordinate with the O atoms featuring 3 distinct coordination geometries, 4-, 5- and 6-fold with Cu-O bonds in the range 1.884 – 2.429 Å that act as magnetic exchange pathways.

4 Magnetic and specific heat studies

Magnetometry data were measured on a Quantum Design Dynacool-9T PPMS system equipped with a VSM option. Two samples were analised: a small crystal glued in a quartz sample holder and a powder sample inside a plastic sample holder, obtained by crushing a polycrystalline piece. Zero-field and field cooling (ZFC, FC) thermomagnetic curves were measured in the range 1.8 – 300 K for applied magnetic fields comprised between 50 Oe and 90 kOe (Fig. 2). A detailed study of the magnetisation process was performed at selected temperatures in the range 1.8 K to 30 K (Fig. 4).

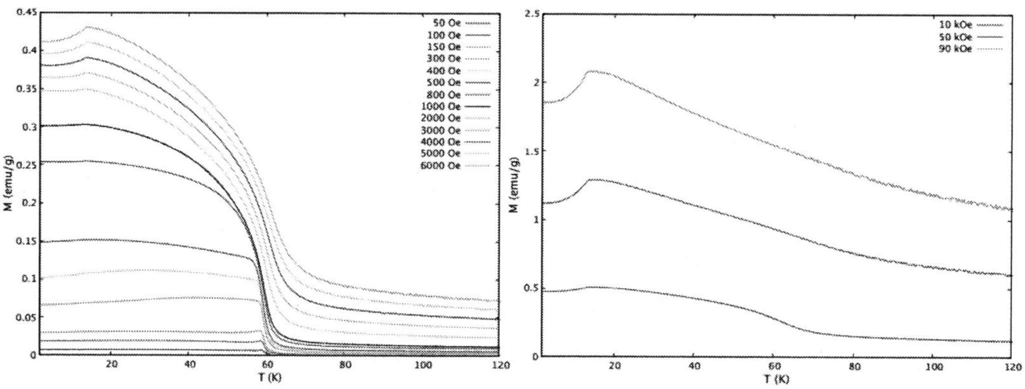

Figure 2. Zero-Field Cooling (ZFC) *M*(*T*) curves for a single crystal of $Cu_4O(SeO_3)_3$ for a set of applied magnetic fields (left panel: low fields; right panel: high fields).

The *M*(*T*) curves (Fig. 2) show a main magnetic transition at 58 K that gets broader with increasing applied field. In addition, a transition at a lower temperature (13 K) was found, corresponding to an anomaly that is clearly visible above ~ 2 kOe. The anomaly becomes more prominent in increasing applied fields. To better characterise this transition, detailed *M*(*H*) curves were measured in the range 1.8–30 K up to the highest available field (90 kOe). Measurements in powder show that in the paramagnetic region the susceptibility follows a modified Curie-Weiss law ($\chi = \chi_0 + C/(T - \Theta)$) with a Curie-Weiss constant corresponding to an effective magnetic moment of 1.84 μ_B/Cu atom, close to that of Cu^{2+} ion, and $\Theta = -81(1)$ K, showing the prevalence of antiferromagnetic interactions (Fig. 3).

The low-temperature *M*(*H*) curves (Fig. 4) could be interpreted as a transition occurring at H_{c1} ~ 400 Oe at 1.8 K shifting to a lower value of ~ 280 Oe at 30 K, likely from an helical into an intermediate conical phase, and a second transition at H_{c2} ~ 1 kOe, above which the net moment increases linearly with applied field. The magnetisation moment value at 90 kOe field and 1.8 K is 0.053 μ_B/Cu, peaking at ~ 0.06 μ_B/Cu at 13 K, far from the full Cu^{2+} moment [4]. Notably, above 13 K and below 30 K a very small remnant is observed in the hysteresis loops.

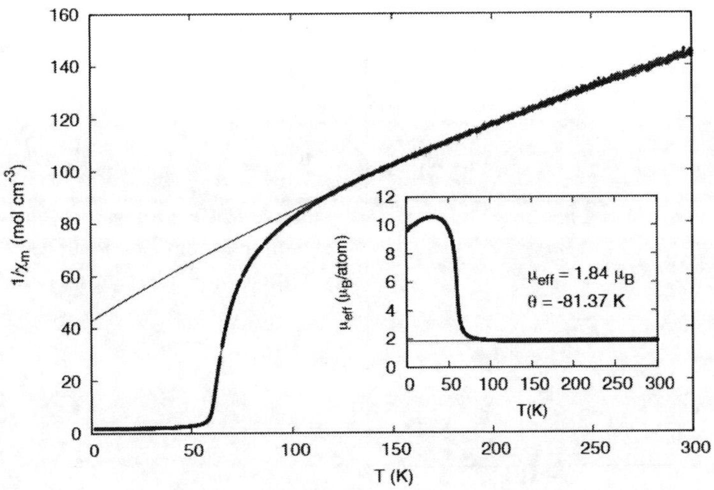

Figure 3. Reciprocal molar susceptibility in a polycrystalline sample measured with an applied field H = 1 kOe, as a function of temperature. The fit of a modified Curie-Weiss law to the data is shown as a solid line. The insert shows the effective magnetic moment as function of temperature.

The obtained results for monoclinic $Cu_4O(SeO_3)_3$ show resemblances to those of the topological chiral magnet Cu_2OSeO_3, with an additional transition at 13 K. However, the curves in Fig. 4 are still far from saturation at 90 kOe, in contrast to the case of Cu_2OSeO_3.

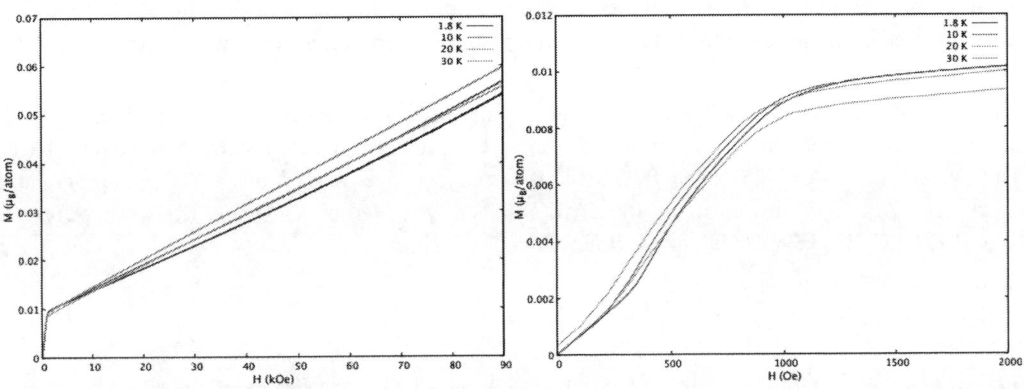

Figure 4. Magnetic-field dependence of the magnetisation ($M(H)$) at low temperature. The right panel shows in detail the low field region.

The same small $Cu_4O(SeO_3)_3$ single crystal used for magnetic measurements was also used for specific heat measurements, under different applied magnetic fields up to 90 kOe, using a semi-adiabatic thermal relaxation method as implemented in the PPMS specific-heat option from Quantum Design. The transition at 13 K found in the magnetometric data was also present in the specific heat data as a clear anomaly found at the same temperature. A plot of $C_p(T)$ is depicted in Fig 5. The transition slightly shifts to lower temperatures as the applied magnetic field is increased. This behaviour is consistent with the main antiferromagnetic character of the magnetic ground state.

5 Conclusions

Monoclinic $Cu_4O(SeO_3)_3$ has a complex structure with a large number of symmetry independent atoms, capable of hosting complex magnetic structures. The magnetic behaviour of

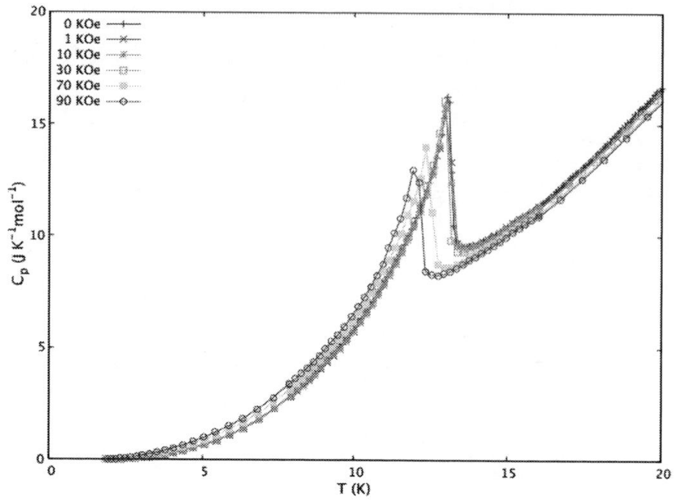

Figure 5. Heat Capacity (C_p) as function of temperature under different applied magnetic fields.

the monoclinic polymorph of $Cu_4O(SeO_3)_3$ shows clear resemblance to that of Cu_2OSeO_3, reported in literature. It is an antiferromagnet with a Néel temperature similar to that of Cu_2OSeO_3 but it shows an additional transition at 13 K, observed both in magnetometric as well as in specific heat measurements. Under an applied magnetic field, the likely helimagnetic antiferromagnetic ground state acquires a small ferromagnetic component.

This work was partially supported by funds from FEDER (Programa Operacional Factores de Competitividade COMPETE) and from FCT under the projects UID/FIS/04564/2016, UID/MULTI/04349/2013, and PTDC/FIS-NAN/6099/2014. JFM acknowledges a grant from FCT PhD programme ChemMat. Access to instruments of the TAIL-UC facility funded under QREN-Mais Centro project ICT_2009_02_012_1890 is gratefully acknowledged as well.

References

[1] S. Seki, X.Z. Yu, S. Ishiwata, Y. Tokura, Science **336**, 198 (2012)

[2] P. Milde, D. Kohler, J. Seidel, L.M. Eng, A. Bauer, A. Chacon, J. Kindervater, S. Muhlbauer, C. Pfleiderer, S. Buhrandt et al., Science **340**, 1076 (2013)

[3] R.B. Versteeg, I. Vergara, S.D. Schäfer, D. Bischoff, A. Aqeel, T. Palstra, M. T.M. Grüninger, P.H. Van Loosdrecht, Physical Review B **94**, 1 (2016)

[4] E. Ruff, P. Lunkenheimer, A. Loidl, H. Berger, S. Krohns, Nature: Scientific Reports **5**, 15025 (2015)

[5] H. Effenberger, F. Pertlik, Monatshefte fuir Chemie **117**, 887 (1986)

[6] N. Kanari, E. Allain, S. Shallari, F. Diot, S. Diliberto, F. Patisson, J. Yvon, Materials **12**, 1625 (2019)

[7] J.R. Panella, B.A. Trump, G.G. Marcus, T.M. McQueen, Crystal Growth and Design **17**, 4944 (2017)

[8] J.F. Malta, M.S.C. Henriques, J.A. Paixão, A.P. Gonçalves, Journal of Magnetism and Magnetic Materials **474**, 122 (2019)

Lattice-Boltzmann simulation of free nematic-isotropic interfaces

Rodrigo C. V. Coelho[1,2,*], *Nuno A. M.* Araújo[1,2], and *Margarida M.* Telo da Gama[1,2]

[1]Centro de Física Teórica e Computacional, Faculdade de Ciências, Universidade de Lisboa, P-1749-016 Lisboa, Portugal.
[2]Departamento de Física, Faculdade de Ciências, Universidade de Lisboa, P-1749-016 Lisboa, Portugal.

Abstract. We use a hybrid method of lattice Boltzmann and finite differences to simulate flat and curved interfaces between the nematic and isotropic phases of a liquid crystal described by the Landau-de Gennes theory. For the flat interface, we measure the interfacial velocity at different temperatures around the coexistence. We show that the interface is completely static at the coexistence temperature and that the profile width is in line with the theoretical predictions. The interface is stable in a range of temperatures around coexistence and disappears when one of the two phases becomes mechanically unstable. We stabilize circular nematic domains by a shift in temperature, related to the Laplace pressure, and estimate the spurious velocities of these lattice Boltzmann simulations.

1 Introduction

Liquid crystals are systems that flow as liquids but where the particles are partially organized in crystal-like ways [1, 2]. They are formed by elongated particles the orientations of which may point in a preferential direction. At low temperatures, these liquid crystals are in the nematic phase (orientationally ordered) while at high temperatures, they are in the isotropic phase, where the orientation of the particles is random (orientationally disordered). At the nematic-isotropic temperature, the two phases coexist and a stable interface may be observed. The standard theory that describes this static interface is based on a tensor order parameter expansion, with gradient terms that account for the order parameter variations, proposed by de Gennes [1].

To simulate the dynamics of liquid crystals, it is common to solve numerically the continuum equations of Beris-Edwards and Navier-Stokes [3–5]. A hybrid method of lattice Boltzmann [6] and finite differences [7] is used. Special care has to be taken in order to simulate interfaces due to spurious currents that arise in these numerical methods, which can lead to interfacial motion or changes in the director field [6, 8].

We analyze the applicability of the method of Ref. [4] to simulate free interfaces, which are very sensitive to these spurious numerical effects. We show that it is possible to simulate static flat interfaces at coexistence, with the profiles predicted by the Landau-de Gennes theory. We estimate the magnitude of the spurious velocities at nearly static circular interfaces stabilized by temperature shifts, related to the Laplace pressure arising from the curvature.

*e-mail: rcvcoelho@fc.ul.pt

1.1 Equations of motion

To simulate the liquid crystal dynamics, we solve the Beris-Edwards equation coupled to the Navier-Stokes equation. The first describes the evolution of the order parameter $Q_{\alpha\beta} = S(n_\alpha n_\beta - \delta_{\alpha\beta}/3)$

$$\partial_t Q_{\alpha\beta} + u_\gamma \partial_\gamma Q_{\alpha\beta} - S_{\alpha\beta}(W_{\alpha\beta}, Q_{\alpha\beta}) = \Gamma H_{\alpha\beta}. \tag{1}$$

Here, ξ stands for the alignment parameter, Γ is the rotational diffusive constant and the co-rotational term is given by:

$$S_{\alpha\beta} =(\xi D_{\alpha\gamma} + W_{\alpha\gamma})\left(Q_{\beta\gamma} + \frac{\delta_{\beta\gamma}}{3}\right) + \left(Q_{\alpha\gamma} + \frac{\delta_{\alpha\gamma}}{3}\right)(\xi D_{\gamma\beta} - W_{\gamma\beta}) - 2\xi\left(Q_{\alpha\beta} + \frac{\delta_{\alpha\beta}}{3}\right)(Q_{\gamma\epsilon}\partial_\gamma u_\epsilon), \tag{2}$$

where $W_{\alpha\beta} = (\partial_\beta u_\alpha - \partial_\alpha u_\beta)/2$, $D_{\alpha\beta} = (\partial_\beta u_\alpha + \partial_\alpha u_\beta)/2$. The Landau-de-Gennes free energy reads

$$\mathcal{F} = \int_V d^3r \Big[\frac{A_0}{2}\left(1 - \frac{\gamma}{3}\right) Q_{\alpha\beta}Q_{\alpha\beta} - \frac{A_0\gamma}{3} Q_{\alpha\beta}Q_{\beta\gamma}Q_{\gamma\alpha} + \frac{A_0\gamma}{4}(Q_{\alpha\beta}Q_{\alpha\beta})^2 + \frac{K}{2}(\partial_\gamma Q_{\alpha\beta})(\partial_\gamma Q_{\alpha\beta})\Big], \tag{3}$$

where A_0 is a constant, γ is a temperature related parameter and K is the elastic constant. Thus, the molecular field, $H_{\alpha\beta} = -\delta\mathcal{F}/\delta Q_{\alpha\beta} + (\delta_{\alpha\beta}/3)\,\mathrm{Tr}(\delta\mathcal{F}/\delta Q_{\gamma\epsilon})$, is

$$H_{\alpha\beta} = -A_0\left(1 - \frac{\gamma}{3}\right) Q_{\alpha\beta} - A_0\gamma\left(Q^2_{\nu\gamma}\frac{\delta_{\alpha\beta}}{3} - Q_{\alpha\gamma}Q_{\beta\gamma}\right) - A_0\gamma Q_{\alpha\beta}Q^2_{\gamma\lambda} + K\partial_\gamma\partial_\gamma Q_{\alpha\beta}. \tag{4}$$

The continuity and Navier-Stokes equations describe the evolution of the density and velocity of the fluid

$$\partial_t\rho + \partial_\alpha(\rho u_\alpha) = 0, \qquad \rho\partial_t u_\alpha + \rho u_\beta\partial_\beta u_\alpha = \partial_\beta\Pi_{\alpha\beta} + \eta\partial_\beta\left(\partial_\alpha u_\beta + \partial_\beta u_\alpha\right). \tag{5}$$

where, the stress-tensor is

$$\begin{aligned}\Pi_{\alpha\beta} =& - P_0\delta_{\alpha\beta} + 2\xi\left(Q_{\alpha\beta} + \frac{\delta_{\alpha\beta}}{3}\right)Q_{\gamma\epsilon}H_{\gamma\epsilon} - \xi H_{\alpha\gamma}\left(Q_{\gamma\beta} + \frac{\delta_{\gamma\beta}}{3}\right) - \xi\left(Q_{\alpha\gamma} + \frac{\delta_{\alpha\gamma}}{3}\right)H_{\gamma\beta} \\ &- \partial_\alpha Q_{\gamma\nu}\frac{\delta\mathcal{F}}{\delta(\partial_\beta Q_{\gamma\nu})} + Q_{\alpha\gamma}H_{\gamma\beta} - H_{\alpha\gamma}Q_{\gamma\beta}.\end{aligned} \tag{6}$$

$P_0 = \rho c_s^2$ is the hydrostatic pressure. Using Eq. (3), we obtain $\delta\mathcal{F}/\delta(\partial_\beta Q_{\gamma\nu}) = K\partial_\beta Q_{\gamma\nu}$, which can be replaced in Eq [**?**].

1.2 Hybrid method

Our numerical scheme solves Eq. (1) using finite differences and Eq. (5) using lattice Boltzmann. Both methods are solved using the same grid (spatial discretization) and the same time step.

1.2.1 Lattice Boltzmann

The lattice Boltzmann method is a numerical technique which solves the Boltzmann equation and recovers the Navier-Stokes equation in the macroscopic limit. The space is discretized in a regular grid and the velocity space is discretized according to the lattice (Gaussian quadrature). Here, we use the D3Q19 lattice, which has 19 velocity vectors $\mathbf{c}_i$ isotropically distributed in three dimensions. The discrete version of the Boltzmann equation is implemented in the method

$$f_i(\mathbf{x} + \mathbf{c}_i\Delta t, t + \Delta t) - f_i(\mathbf{x}, t) = (\partial f_i/\partial t)_{\mathrm{coll}} + S_i, \tag{7}$$

where f_i is the distribution function corresponding to the $\mathbf{c}_i$ vector. The simplest collision operator is the Bhatnagar–Gross–Krook (BGK) one, which assumes that the full-equilibrium distribution f_i relaxes to the equilibrium f_i^{eq} with a characteristic relaxation time τ: $(\partial f_i/\partial t)_{\text{coll}} = -(\Delta t/\tau)(f - f^{eq})$. The multi-relaxation time operator, used in this work, is a generalization of the BGK: $(\partial f_i/\partial t)_{\text{coll}} = \mathbf{M}^{-1}\mathbf{R}\mathbf{M}[f_i(\mathbf{x}, t) - f_i^{eq}(\mathbf{x}, t)]\Delta t$, where the matrix $\mathbf{M}$ transforms from the distribution function space to the hydrodynamic moments space and $\mathbf{R}$ is the relaxation matrix. It assumes that the hydrodynamic moments may relax with different relaxation times. This approach is known to improve the accuracy and stability in many problems [6]. The equilibrium distribution is the second order expansion in Hermite polynomials of the Maxwell-Boltzmann distribution [9]:

$$f_i^{eq} = \rho w_i \left[1 + \frac{\mathbf{c}_i \cdot \mathbf{u}}{c_s^2} + \frac{(\mathbf{c}_i \cdot \mathbf{u})^2}{2c_s^4} - \frac{\mathbf{u}^2}{2c_s^2} \right], \tag{8}$$

where c_s is the speed of sound in the given lattice. For the D3Q19, it is $c_s = 1/\sqrt{3}$. The density and velocity fields are calculated as follows:

$$\rho = \sum_i f_i, \quad \rho\mathbf{u} = \sum_i \mathbf{c}_i f_i + \frac{\mathbf{F}_i \Delta t}{2}. \tag{9}$$

The source term, for the BGK operator, reads: $S_i = w_i\left(1 - \frac{1}{2\tau}\right)\left[\frac{\mathbf{c}_i}{c_s^2}\left(1 + \frac{\mathbf{c}_i \cdot \mathbf{u}}{c_s^2}\right) - \frac{\mathbf{u}}{c_s^2}\right] \cdot \mathbf{F}$. For the MRT operator, this source term must be relaxed in the moments space as the distribution function (see Refs. [4, 6] for more details of the MRT method). The force is calculated using the stress tensor of Eq. (6): $F_\alpha = \partial_\beta(\Pi_{\alpha\beta} + P_0\delta_{\alpha\beta})$.

1.2.2 Finite differences

The Eq. (1) is solved explicitly using a predictor-corrector finite difference method. All the differences are second order accurate. For instance, the first derivative of a vector V in the x-direction is: $\partial_x V(x) = [V(x+\Delta x) - V(x-\Delta x)]/(2\Delta x) + O(\Delta x^2)$. To calculate the time evolution of $Q_{\alpha\beta}$, we follow the following steps: 1) Calculate the time derivative, $(\partial Q_{\alpha\beta}/\partial t)'$, using Eq. (1) with the current $Q_{\alpha\beta}$; 2) Calculate the predictor: $Q^P_{\alpha\beta} = Q_{\alpha\beta} + \Delta t(\partial Q_{\alpha\beta}/\partial t)'$; 3) Calculate the time derivative using the predictor, $(\partial Q^P_{\alpha\beta}/\partial t)$, with Eq. (1); 4) Calculate the corrected derivative: $\frac{\partial Q_{\alpha\beta}}{\partial t} = \frac{1}{2}\left[\left(\frac{\partial Q_{\alpha\beta}}{\partial t}\right)' + \left(\frac{\partial Q^P_{\alpha\beta}}{\partial t}\right)\right]$; 5) Calculate the corrector: $Q_{\alpha\beta}(t+\Delta t) = Q_{\alpha\beta}(t) + \Delta t \frac{\partial Q_{\alpha\beta}}{\partial t}$; 6) Return to step 1. Notice that the velocity used in Eq. (1) is the actual one (and not the lattice Boltzmann one: $\rho\mathbf{u}_{LB} = \sum_i \mathbf{c}_i f_i$), given by Eq. (9), which is second order accurate.

2 Free interface

In this section, we show the results from numerical simulations of unconfined nematic-isotropic interfaces. First, we consider a flat interface in two dimensions and then we simulate a circular interface. In the simulations, we apply periodic boundary conditions in both directions. The initial velocity is set to zero and the density to $\rho = 1$ everywhere. Other parameters are: $\tau = 1.5$, $K = 0.04$, $\xi = 0.7$, $A_0 = 0.1$ and $\Gamma = 0.34$. Our results are given in lattice units: the distance between nodes is $\Delta x = 1$ and the time step is $\Delta t = 1$.

2.1 Flat interface

In order to apply periodic conditions, we initialize our system as follows: the nematic phase, with directors in the vertical direction, is in the center of the domain with isotropic phase

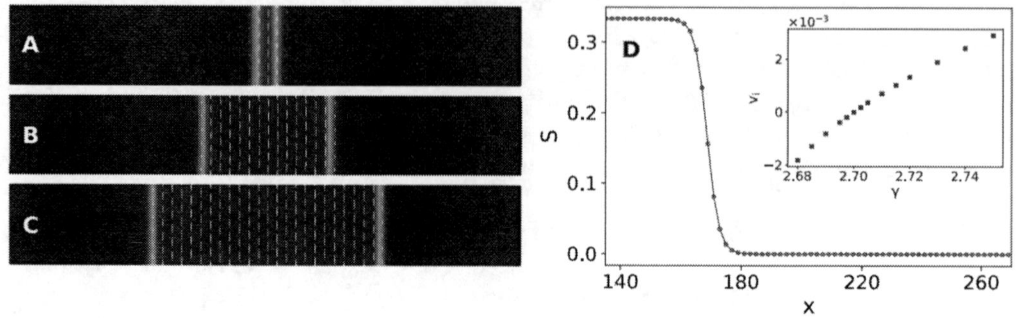

Figure 1. Order parameter for flat interfaces at different temperatures and at time $t = 350000$: A) $\gamma = 2.699$, B) $\gamma = 2.7$, C) $\gamma = 2.701$. The velocity field is below the machine precision everywhere for $\gamma = 2.7$. D) Scalar order parameter at $y = L_Y/2$. The solid line is the analytical expression given by Eq. (10) for $S_n = 1/3$ and $\lambda = 2$ and the circles are the simulated results. The inset shows the interface velocity at different temperatures. At coexistence $\gamma = 2.7$, the interface velocity is below the machine precision.

elsewhere. Thus, we simulate two interfaces, but we focus our analysis on the right interface. At the coexistence temperature ($\gamma_{NI} = 2.7$) the interface is completely static and there are no spurious velocities. Eq. (3) allows an interface solution of the form [10]

$$S = \frac{S_n}{2}\left[1 - \tanh\left(\frac{x - x_i}{2\lambda}\right)\right], \tag{10}$$

where S_n is the nematic order parameter and $\lambda = \sqrt{27K/A_0\gamma}$ is the correlation length. At $\gamma = 2.7$, $\lambda = 2$ and $S_n = 1/3$. Fig. 1 D shows that the simulation reproduces accurately the interfacial profile.

Next, we change the temperature slightly away from coexistence. In this case, the interface moves with constant velocity, which can be to the left (negative) or to the right (positive). Figs. 1 A and C, show the interfaces at two different temperatures after an interval of time. For $\gamma < \gamma_{NI}$ the domain of the nematic phase shrinks and for $\gamma > \gamma_{NI}$ the nematic domain expands. In the inset of Fig. 1D, we plot the interface velocity at different temperatures. It is approximately linear close to γ_{NI} but is deviates from the linear behavior as the shift from coexistence increases. Above a certain shift from γ_{NI} one of the two phases becomes unstable and the entire domain becomes nematic or isotropic in the absence of interfacial propagation. Below $\gamma_< = 2.65$ the system becomes isotropic and above $\gamma_> = 3.1$ it becomes nematic.

2.2 Circular interface and spurious velocities

Here we estimate the spurious velocities generated at a circular interface. The spurious velocities arise at curved interfaces as a result of the discretization of the velocity space. However, it is difficult to distinguish the spurious velocities from the physical ones in this multiphase model, because curved interfaces are usually unstable. Fig. 2E shows the interface position at three different temperatures. In order to identify the spurious velocities, we iteratively search for the temperature where the interface velociy is of the order of the spurious ones. We found that for $\gamma = 2.707681$ the interface velocity at $t = 35000$ is approximately $u_i \approx 1.89 \times 10^{-7}$. Fig. 2A and B show the oder parameter and velocity field of this almost static interface. From the vorticity field, we notice that the there are vortices between the directions of the

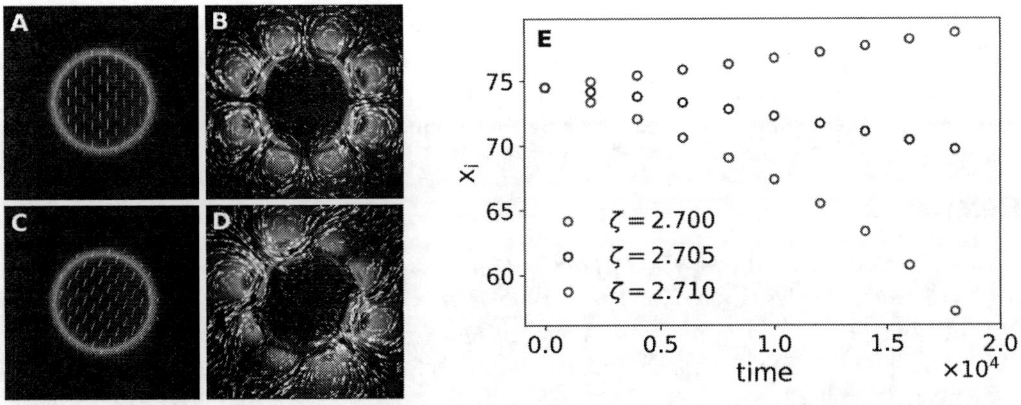

Figure 2. Order and velocity fields of an almost static circular interface at two different director orientations. The temperature, $\gamma = 2.707681$ was chosen to minimize the interfacial velocity and the screenshots were taken at $t = 35000$. The system size is 100×100, the radius of the circle is $R = 25$. A) Order parameter (colors) and directors (lines) for $n_y = 1$ and $n_x = 0$. B) Vorticity field (colors) and direction of the velocity field (arrows). C and D depict the same fields as A and B, but for $n_y = \cos(20°)$ and $n_x = \sin(20°)$. In this color map, the red and blue represent the maximum and minimum values of the field. The maximum velocity (norm) in B is $u_{max} = 8.97 \times 10^{-7}$ and $u_{max} = 1.13 \times 10^{-6}$ in D. E) Interface velocity at three different temperatures.

velocity vectors of the D3Q19 lattice, which is an indication that these are spurious currents. In Fig. 2 D we plot the velocity field for the same system rotated by 20°. One notes that the intensity of the vortices changed due to the physical component of these velocities (the small interface velocity), but the positions of the vortices are the same confirming that they are mostly due to the velocity discretization. From this analysis, we estimate that the spurious velocities are around 10^{-6} in lattice units for this setup. If necessary, one could reduce the spurious velocities by increasing the isotropy of the lattice [11] or using more advanced collision operators [12].

3 Conclusions

We examined the applicability of the hybrid method of lattice Boltzmann and finite differences for the simulation of nematic-isotropic interfaces. The shape and width of the interface is correctly reproduced and the free interface is static at the coexistence temperature. At temperatures slightly different from the coexistence, the interface moves with constant velocity that depends on the temperature. The interface exists only for a certain range of temperatures: for temperatures much lower or higher than the coexistence, one of the two phases becomes unstable and the interface disappears without propagation. The circular interfaces are not static at the coexistence temperature. Thus, we stabilize them by a temperature shift in a way that the interface velocity is comparable to the spurious velocities. This allows us to visualize the profile of theses spurious currents and to estimate their magnitude.

Acknowledgements

We acknowledge financial support from the Portuguese Foundation for Science and Technology (FCT) under the contracts: PTDC/FIS-MAC/28146/2017 (LISBOA-01-0145-FEDER-028146) and UID/FIS/00618/2019. Margarida Telo da Gama (MTG) would like to thank the

Isaac Newton Institute for Mathematical Sciences for support and hospitality during the program "The mathematical design of new materials" where most of this work was carried out. This program was supported by EP-SRC Grant Number: EP/R014604/1. MTG participation in the program was supported in part by a Simons Foundation Fellowship.

References

[1] P.G. de Gennes, J. Prost, *The Physics of Liquid Crystals*, International Series of Monographs on Physics (Clarendon Press, 1995), ISBN 9780198517856

[2] A. Beris, B. Edwards, *Thermodynamics of Flowing Systems: with Internal Microstructure*, Oxford Engineering Science Series (Oxford University Press, 1994), ISBN 9780195344882

[3] D. Marenduzzo, E. Orlandini, M.E. Cates, J.M. Yeomans, Physical Review E **76**, 031921 (2007)

[4] R.C.V. Coelho, N.A.M. Araújo, M.M. Telo da Gama, Soft Matter (2019), accepted manuscript - DOI:10.1039/C9SM00859D

[5] A. Doostmohammadi, M.F. Adamer, S.P. Thampi, J.M. Yeomans, Nature Communications **7**, 10557 (2016)

[6] T. Krüger, H. Kusumaatmaja, A. Kuzmin, O. Shardt, G. Silva, E.M. Viggen, *The Lattice Boltzmann Method - Principles and Practice* (Springer International Publishing, 2016), ISBN 978-3-319-44647-9

[7] F. Vesely, *Computational Physics: An Introduction* (Springer US, 2001), ISBN 9780306466311, `https://www.springer.com/gp/book/9780306466311`

[8] Z. Yu, L.S. Fan, Physical Review E **82**, 046708 (2010)

[9] R.C.V. Coelho, M.M. Doria, Computers & Fluids **165**, 144 (2018)

[10] O.N. Mesquita, Brazilian Journal of Physics **28**, 04 (1998)

[11] X. Shan, Physical Review E **73**, 047701 (2006)

[12] C. Coreixas, B. Chopard, J. Latt, arXiv e-prints arXiv:1904.12948 (2019), `1904.12948`

Theoretical calculations of nonlinear optical calculations of 2D materials

G. B. Ventura[1,*], *D. J.* Passos[1], *J. M.* Viana Parente Lopes[1], and *J. M. B.* Lopes dos Santos[1]

[1]Centro de Física das Universidades do Minho e Porto, Universidade do Porto, 4169-007, Porto, Portugal

Abstract. One important feature of two dimensional (2D) materials is that they possess an exceptional nonlinear optical (NLO) response to light, with conductivities that are several orders of magnitude larger than their 3D counterparts. The theoretical descriptions of these NLO responses in crystalline systems involve two different representations of the perturbation: the length and velocity gauges. The former has been the formalism of choice for the past two decades; the latter was implemented only recently, due to concerns that it could not be pratically implemented without breaking sum rules – a set of identities that ensure the equivalence between the two formalisms – which would then render the results unphysical. In this work, we shall review and summarize our contributions to the study of the two formalisms and of their relationship by means of the aforementioned sum rules.

Nonlinear optical responses

The description of the nonlinear optical response of electrons in a crystal to an external, spatially constant and monochromatic, electric field was first formulated – in terms of the density matrix formalism – in Claudio Aversa and J. E. Sipe's work of 1995 [1]. There, it is considered that, in the absence of the electric field perturbation, an electron in a crystal can described by the periodic Hamiltonian, $H_0(\mathbf{k})$, of eigenvalues (bands), $\epsilon_{\mathbf{k}s}$, and eigenvectors (Bloch states), $\Psi_{\mathbf{k}s}$, both of which are labelled by a crystal momentum $\mathbf{k}$, a vector in the first Brillouin zone (FBZ) and a band index, s. Electrons are deemed to be non-interacting among themselves – that is, they are subjected only to the electric field of light that is shining on the material – and so the many-particle ground state is the Fermi sea, a state whose occupation numbers follow from the Fermi-Dirac distribution function, $\langle c^{\dagger}_{\mathbf{k}s} c_{\mathbf{k}s'} \rangle = \delta_{ss'} f_{\mathbf{k}s}$. In rather simplistic terms, the effect that the external electric field has on this system of electrons in a set of bands is to change the populations within each band and to introduce transitions between them so that the matrix elements of the density matrix (DM), $\rho_{\mathbf{k}ss'}(t) = \langle c^{\dagger}_{\mathbf{k}s}(t) c_{\mathbf{k}s'}(t) \rangle$, are now dynamical. The equations of motion for the DM matrix elements are given by,

$$i\hbar\dot{\rho}_{\mathbf{k}ss'}(t) = (\epsilon_{\mathbf{k}s} - \epsilon_{\mathbf{k}s'})\rho_{\mathbf{k}ss'}(t) + [V(t), \rho(t)]_{\mathbf{k}ss'}. \quad (1)$$

A perturbative treatment of $\rho_{\mathbf{k}ss'}(t)$ then allows us to study the crystal's response functions – here we shall consider the conductivity – in its different orders of the electric field [2]. To

*e-mail: gbventura@fc.up.pt

do so, one must first specify the form of the perturbation, $V(t)$. Due to the inherent gauge freedom involved in expressing the electric field in terms of the scalar and vector potentials, there is no unique way of writing the perturbation; for simplicity, either the vector potential or the scalar potential is set to zero, and obtains the length and velocity gauges, respectively:

$$V^E(t) = e\mathbf{E}(t) \cdot \mathbf{r}; \tag{2}$$

$$V^A(t) = H_0(\mathbf{k} + e\hbar^{-1}\mathbf{A}(t)) - H_0(\mathbf{k}). \tag{3}$$

Similarly to electromagnetism, the choice of a gauge – we shall presently see – has its own particular advantages and disadvantages. We also note that the equivalence between the descriptions associated to each gauge, though following directly from gauge invariance, has to be reconsidered in light of practical calculations.

Length gauge

In the length gauge, the perturbation is expressed in terms of the position operator and, as such, it is not invariant under lattice translations. Some attention is thus needed when expressing this object in terms of Bloch states [3],

$$\mathbf{r}_{\mathbf{k}\mathbf{k}'ss'} = \delta_{ss'}\,(2\pi)^d\,\nabla_{\mathbf{k}}\delta(\mathbf{k}-\mathbf{k}') - i(2\pi)^d\,\delta(\mathbf{k}-\mathbf{k}')\xi_{\mathbf{k}ss'}, \tag{4}$$

for $\xi_{\mathbf{k}ss'}$, the Berry connection [1]. Two equivalent formulations using this gauge have been put forth. In [1], the authors separate the perturbation into intraband $(i : s' = s)$ and interband $(e : s' \neq s)$ portions so that the response at a given order is obtained by considering all relevant possible pairings of the two parts of the perturbation, e.g., in second order, one of the terms reads as ρ_{ie}. In our first work on this subject [4], no such distinction is made: the position operator in the Bloch representation is treated as a covariant derivative, $\mathbf{D}_{\mathbf{k}ss'} = \delta_{ss'}\nabla_{\mathbf{k}} - i\xi_{\mathbf{k}ss'}$. This formulation makes it clear that the Bloch representation of the position matrix element is gauge covariant under phase shifts of the Bloch states. Following [4], the equation of motion for the DM reads as,

$$i\hbar\dot{\rho}_{\mathbf{k}ss'}(t) = (\epsilon_{\mathbf{k}s} - \epsilon_{\mathbf{k}s'})\rho_{\mathbf{k}ss'}(t) + ie\mathbf{E}(t) \cdot [\mathbf{D}, \rho(t)]_{\mathbf{k}ss'}. \tag{5}$$

By expanding $\rho_{\mathbf{k}ss'}(t)$ in powers of the external field, one can see the two main features of this gauge: the complexity of the solutions increases for high order responses, as an nth-order solution involves the $\mathbf{k}$-space derivative of the $(n-1)$th-order one, which is (in principle) a rather complicated function in that variable; in the frequency domain, one can see that energy dependences, $\Delta\epsilon_{\mathbf{k}ss'} = \epsilon_{\mathbf{k}s} - \epsilon_{\mathbf{k}s'}$, appear only in the denominators. Practical calculations in this gauge can then use a truncation of the Hilbert space, both in the number of bands selected and the region of $\mathbf{k}$-space that is taken into consideration, which makes length gauge the most suitable gauge for analytical work.

Velocity gauge

In the case of the velocity gauge, the interaction with the external electric field is introduced via minimum coupling. This means that the form of the associated perturbation, $V(t)$, depends on the Hamiltonian describing the unperturbed system, Eq.(3). So, while the perturbation in the length gauge remains unchanged – whether one is considering the continuum Hamiltonian or a tight-binding one – the same is not true in the velocity gauge, where the perturbation has to be considered on a case-by-case basis. To illustrate a second point concerning the velocity gauge, we consider the continuum Hamiltonian, $H_0(\mathbf{k}) = (\hbar^2/2m)(-i\nabla + \mathbf{k})^2 + U(\mathbf{r})$, for $U(\mathbf{r})$ the periodic lattice potential. The corresponding perturbation, $V^A(t)$, written in the Bloch basis, reads as,

$$V^A(t)_{\mathbf{k}ss'} = e\mathbf{A}(t) \cdot \mathbf{v}_{\mathbf{k}ss'} + \delta_{ss'} \frac{e^2}{2m} \mathbf{A}(t) \cdot \mathbf{A}(t). \tag{6}$$

The relevant portion of the perturbation is proportional to the velocity operator, which being lattice periodical, can be easily expressed in terms of Bloch states,

$$\mathbf{v}_{\mathbf{k}ss'} = \delta_{ss'} \nabla_{\mathbf{k}} \epsilon_{\mathbf{k}s} - i\Delta\epsilon_{\mathbf{k}ss'} \xi_{\mathbf{k}ss'}. \tag{7}$$

One can easily see that for such a perturbation, the solutions to the equations of motion for the DM will have an energy dependence on $\Delta\epsilon_{\mathbf{k}ss'}$ in the numerators (as well as in the denominators) since the interband matrix elements of the velocity carry such energy factors. The velocity gauge, therefore, does not allow for a truncation of the Hilbert space along the same lines as the length gauge and so practical calculations in the former gauge have to be implemented differently than those in the latter. This sticking point was only fully solved in our second work [5] and required a closer study on how equivalence is ensured in these calculations.

Equivalence between procedures and sum rules

Gauge invariance tells us that an observable computed in the length gauge must be necessarily equal to its counterpart in the velocity gauge,

$$\langle O_E(t) \rangle = \langle O_A(t) \rangle. \tag{8}$$

In the present case, the operators are related by a unitary transformation that describes time-dependent translations in momentum space. By expressing these operators in the Bloch basis, one can see that Eq.(8) holds if and only if a set of sum rules are fulfilled; these sum rules are a set of integrals that must be shown to be trivial – a condition that is satisfied when the original operators are defined in the full FBZ and if the set of bands considered remains untruncated [4]. By enforcing these criteria from the start, it is possible to formulate a procedure for practical calculations in the velocity gauge whose results are equivalent to those in the length gauge [5]. We must emphasize that this seemingly technical point has important consequences for the calculations of NLO since the velocity gauge – being fully diagonal in $\mathbf{k}$-space – is better suited for numerical implementations.

Numerical results in the velocity gauge: THG in the monolayer of graphene

A proper calculation in the velocity gauge requires the use of an Hamiltonian that is defined in the full FBZ and whose set of bands cannot be truncated. In [5], we compared the results for the third harmonic generation [2] using the velocity gauge description for the standard two band tight-binding model of graphene with those following from a length gauge one. These results are shown here in Figure 1. One can see that the perfect agreement between the two procedures.

Conclusion

For the past twenty years, the length gauge has been the formalism of choice for computing the nonlinear optical response of crystals while its counterpart formalism – the velocity gauge

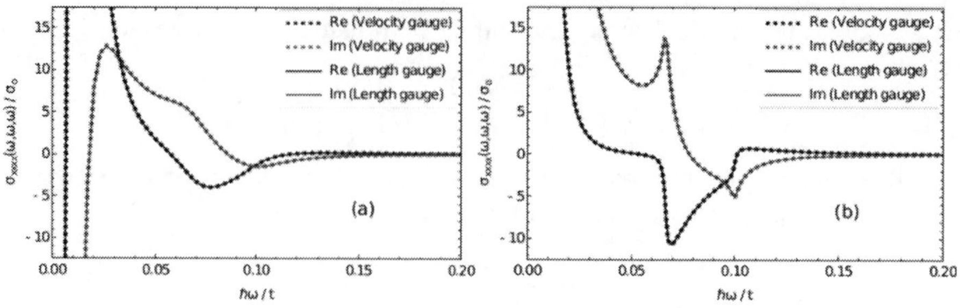

Figure 1. Frequency dependence of the third-order nonlinear conductivity of graphene, normalized to σ_0 at zero temperature. The parameters used here were (a) $\mu/t = 0.1$, $\gamma/t = 0.011$; (b) $\mu/t = 0.1$, $\gamma/t = 0.001$, for γ and t the scattering and hopping parameters, respectively. The solid curves were obtained from a length gauge approach and the dashed ones from a velocity gauge calculation. Image and caption were taken from [5].

– was considered to be of little use for practical calculations. In our recent works [4, 5], we have described a more workable representation of the perturbation of the length gauge; related the observables in the length and velocity gauges; understood how the sum rules that ensure the equivalence between procedures are broken; and more importantly, we developed a working formalism for calculations in the velocity gauge.

Acknowledgments

The authors acknowledge financing of Fundação da Ciência e Tecnologia, of COMPETE 2020 program in FEDER component (European Union), through projects POCI-01-0145-FEDER-028887 and UID/FIS/04650/2013.

References

[1] C. Aversa and J. E. Sipe, Phys. Rev. B **52**, 14636 (1995)
[2] R. W. Boyd, *Nonlinear Optics* (Academic Press, Third Edition, 2008)
[3] E. I. Blount, Formalisms of band theory, Solid State Phys. **13**, 305 (1962)
[4] G. B. Ventura, D. J. Passos, J. M. Viana Parente Lopes, J. M. B. Lopes dos Santos and N. N. R. Peres, Phys. Rev. B **96**, 035431 (2017)
[5] D. J. Passos, G. B. Ventura, J. M. Viana Parente Lopes, J. M. B. Lopes dos Santos and N. N. R. Peres, Phys. Rev. B **97**, 235446 (2018)

Non-linear optical response in disordered 2D materials

Simão M. João[2,*] and *João* M. Viana P. Lopes[2,]

[1]Centro de Física das Universidades do Minho e Porto, Universidade do Porto, 4169-007, Porto, Portugal. Departamento de Física e Astronomia, Faculdade de Ciências da Universidade do Porto.

Abstract. Using KITE [1], a quantum transport software developed by ourselves, we explore the effect of disorder in the second-order conductivity, aiming to reproduce mesoscopic samples under more realistic models of disorder. This work will be concerned about our most recent results with KITE. We will showcase and examine how different models of disorder affect the same system, experimenting with Anderson disorder and vacancies in gapped Graphene.

1 Introduction

Two-dimensional materials are a topic of great interest in the realm of nonlinear optics because they possess some of the strongest nonlinear optical responses to light. The bulk of the theoretical work developed to understand the nonlinear optical response of these materials [2–5] is concerned with translational-invariant systems, where disorder may be added perturbatively. In our work, we extend the work of Passos et al [6, 7] by implementing the velocity gauge methodology within the Keldysh formalism [8]. We developed a general perturbation procedure to deal with non-interacting fermion systems at finite temperature coupled to a time-dependent external electric field. We express the formula for the n-th order optical conductivity in terms of a trace, reflecting the fact that translation invariance was not needed for its derivation.

These expressions gain particular utility when used in the real-space basis, for then disorder may be introduced non-perturbatively and studied in, for example, tight-binding models. This includes Anderson disorder and any kind of structural disorder such as adatoms, vacancies and defects. This generality also allows one to study the effect of a magnetic field (which, through a Peierls phase, breaks translation symmetry on its own) in disordered systems.

Spectral methods such as the Kernel Polynomial Method (KPM) [9] become a very powerful tool to evaluate these traces, which, coupled with these formulas, enable the computation of the second-order response of large (10^6 atoms) disordered systems with a resolution of 10 meV [8]. These formulas, up to second order, are implemented in KITE, an open-source software tailored for scalable and efficient quantum transport simulations [1].

In the first part of this proceeding, we start by presenting the expressions obtained through the Keldysh framework [10]. Then, after a brief explanation on KPM we use

*e-mail: simao.joao@fc.up.pt

this framework to explore different models of disorder in gapped Graphene (GG), one of the simplest models for which we can find a second-order response to an external electric field. Since we are only interested in the intrinsic nonlinear optical response of the material, we are ignoring possible excitonic effects. Although we present GG specifically, KITE is able to calculate the transport properties of a very broad range of materials.

2 Nonlinear optical conductivity with the Keldysh formalism

Using the Keldysh formalism [10], we were able to obtain closed-form expressions for the general n-th order optical conductivity [8]. We only show the formulas up to second-order because the higher-order ones are too computationally expensive to be of practical use. The formulas for the first and second order conductivities obtained are, respectively,

$$\sigma^{\alpha\beta}(\omega) = \frac{ie^2}{\Omega\omega}\int_{-\infty}^{\infty} d\epsilon f(\epsilon)\mathrm{Tr}\Big[\hat{h}^{\alpha\beta}\delta(\epsilon - H_0) \tag{1}$$

$$+\frac{1}{\hbar}\hat{h}^{\alpha}g^{R}(\epsilon/\hbar+\omega)\hat{h}^{\beta}\delta(\epsilon-H_0)+\frac{1}{\hbar}\hat{h}^{\alpha}\delta(\epsilon-H_0)\hat{h}^{\beta}g^{A}(\epsilon/\hbar-\omega)\Big]$$

and

$$\sigma^{\alpha\beta\gamma}(\omega_1,\omega_2) = \frac{1}{\Omega}\frac{e^3}{\omega_1\omega_2}\int_{-\infty}^{\infty} d\epsilon f(\epsilon)\mathrm{Tr}\Big[\frac{1}{2}\hat{h}^{\alpha\beta\gamma}\delta(\epsilon-H) \tag{2}$$

$$+\frac{1}{\hbar}\hat{h}^{\alpha\beta}\delta(\epsilon-H)\hat{h}^{\gamma}g^{A}(\epsilon/\hbar-\omega_2)+\frac{1}{\hbar}\hat{h}^{\alpha\beta}g^{R}(\epsilon/\hbar+\omega_2)\hat{h}^{\gamma}\delta(\epsilon-H)$$

$$+\frac{1}{2\hbar}\hat{h}^{\alpha}g^{R}(\epsilon/\hbar+\omega_1+\omega_2)\hat{h}^{\beta\gamma}\delta(\epsilon-H)+\frac{1}{2\hbar}\hat{h}^{\alpha}\delta(\epsilon-H)\hat{h}^{\beta\gamma}g^{A}(\epsilon/\hbar-\omega_1-\omega_2)$$

$$+\frac{1}{\hbar^2}\hat{h}^{\alpha}g^{R}(\epsilon/\hbar+\omega_1+\omega_2)\hat{h}^{\beta}g^{R}(\epsilon/\hbar+\omega_2)\hat{h}^{\gamma}\delta(\epsilon-H)$$

$$+\frac{1}{\hbar^2}\hat{h}^{\alpha}g^{R}(\epsilon/\hbar+\omega_1)\hat{h}^{\beta}\delta(\epsilon-H)\hat{h}^{\gamma}g^{A}(\epsilon/\hbar-\omega_2)$$

$$+\frac{1}{\hbar^2}\hat{h}^{\alpha}\delta(\epsilon-H)\hat{h}^{\beta}g^{A}(\epsilon/\hbar-\omega_1)\hat{h}^{\gamma}g^{A}(\epsilon/\hbar-\omega_1-\omega_2)\Big]$$

where $\hat{h}^{\alpha_1\cdots\alpha_n} = \frac{1}{(i\hbar)^n}[\hat{r}^{\alpha_1},[\cdots[\hat{r}^{\alpha_n},H_0]]]$, $\hat{\boldsymbol{r}}$ is the position operator and $g^{A,R}$ are the advanced and retarded Green's functions, respectively. If the system presents translation invariance and a $\boldsymbol{k}$-space basis is used, we recover the known formulas for the nonlinear optical conductivity derived through the density matrix formalism or Kubo's formalism [6, 7]. A phenomenological broadening parameter $0^+ \to \gamma$ is added by hand in the Green's functions to mimic the effect of inelastic scattering. In some cases, for pure systems, the conductivity may be infinite, which stems from the fact that there is no dissipation mechanism. Real systems exhibit some disorder, and to reflect this, the abovementioned phenomenological parameter γ is added. Ideally, one should use real disorder, but this would break translation invariance. This will be the topic for the next sections.

Symmetry may cause some of these terms to be zero. In fact, due to centrosymmetry, Graphene displays no second-order response. GG does not have this symmetry

 https://doi.org/10.1051/epjconf/202023303002

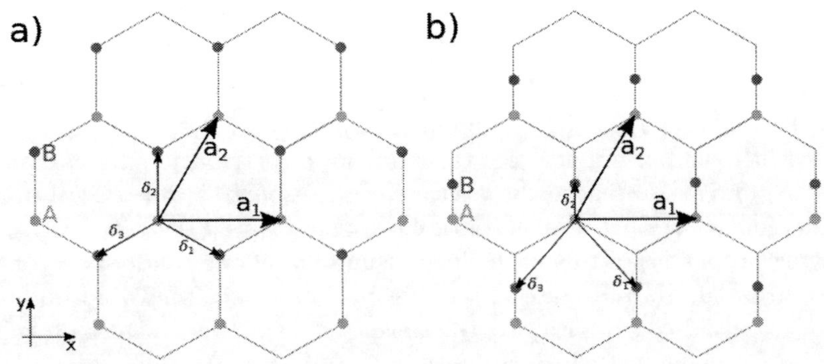

Figure 1. Primitive vectors and atomic basis used in GG (a) and in the related model where one of the sublattices is displaced with respect to the other (b). The B sublattice is displaced by $-\boldsymbol{\delta}_2/2$.

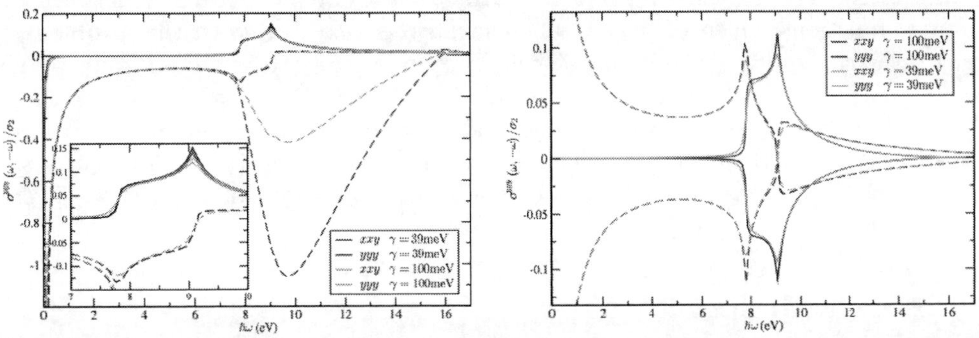

Figure 2. Photogalvanic effect in dGG (left) and GG (right). The gap is $\Delta = 7.78$ eV. Lattice size: $L = 2048$, Fermi energy $\varepsilon_F = 0$ eV. The solid curves represent the real part of the conductivity and the dashed curves the imaginary part.

and therefore has a second-order response. The photogalvanic effect in GG [11] is expected to diverge in the limit $\gamma \to 0^+$ as discussed in the previous paragraph. However, GG still has C_3 rotational symmetry and three reflection axis, and these are in the origin of the cancellation of the terms in eq. 2 responsible for making the conductivity diverge.

To check this, we used the $\boldsymbol{k}$-space basis to perform the calculations for two different values of γ in GG and in another model with less symmetry, which we denote by 'displaced Gapped Graphene' (dGG). This somewhat artificial model is the same as GG, but sublattice B is displaced by $-\boldsymbol{\delta}_2/2$ with respect to sublattice A (see fig. 1), while maintaining the same hopping integrals between neighbours. Therefore, the velocity operator is modified but the Hamiltonian is not. In fig. 2, we see that γ controls the sharpness of the curves in GG, but has otherwise no qualitative effect on the conductivity. In dGG, for the xxy conductivity tensor, the symmetry responsible for making the conductivity diverge is not present, and a clear dependence on γ exists in the imaginary part of the conductivity. For most practical applications, however, this is not problematic since only the real part of the conductivity is relevant.

Eq. 2 is of most value for systems without translation invariance, a topic which is tackled in the following sections.

3 The Kernel Polynomial Method

In order to compute the traces in eqs. 1 and 2 we resort to a Stochastic Trace Evaluation (STE) and the Kernel Polynomial Method (KPM)[9]. Let $|\xi\rangle = \sum_i \xi_i |\boldsymbol{R}_i\rangle$ be a vector with random numbers ξ_i, taken from a distribution X, assigned to each lattice site $\boldsymbol{R}_i$ and $\langle\cdots\rangle$ denote the average with respect to that distribution. Then, imposing $\langle\xi_i^*\xi_j\rangle = \delta_{ij}$ on the distribution, we get $\langle\langle\xi| A |\xi\rangle\rangle = \mathrm{Tr}(A)$. Due to the self-averaging properties of this procedure, usually only one random vector is needed. The functions of the Hamiltonian appearing in eqs. 1 and 2 have no immediate representation in a real-space basis, so they are first expanded in terms of Chebyshev polynomials of the unperturbed Hamiltonian $f(H_0) = \sum_n a_n T_n(H_0)$. Then, the resulting objects to compute are of the form $\langle\xi| T_n(H_0) h^\alpha T_m(H_0) h^\beta |\xi\rangle$. This operation scales linearly with the system size and that is one of the great advantages of KPM. These objects are then multiplied by the respective Chebyshev coefficients of the corresponding expanded functions and summed over the indices. The functions we are interested in expanding are the Dirac delta and the Green's functions, both singular functions which display Gibbs oscillations. To deal with this problem, the function is convoluted with a kernel, which changes the coefficients a_n and removes the singularities. Another way to do this is by adding a finite imaginary part to the Green's function $i0^+ \to i\gamma$, which plays a similar role to the phenomenological scattering parameter. In this case, however, this factor is chosen to be sufficiently small so as to not affect the final results but sufficiently large to obtain a well-converged Green's function. This is the approach we use.

4 Results with disorder

In this section, we explore the effect of different disorder types in GG in the second-order conductivity of this material. Excitonic effects are explicitly ignored since we are only interested in studying the non-interacting part of the conductivity of this material at this point. For this purpose, we use a tight-binding model for GG with only nearest-neighbor hoppings and a gap similar to that of hexagonal Boron-Nitride ($\Delta = 7.8$ eV). The disorder models hereby used are meant to illustrate the capabilities of our method.

4.1 Anderson disorder

The first model of disorder is the simplest - Anderson disorder. Each lattice site is allowed to have a random local energy taken from a uniform distribution in the range $\left[-\frac{W}{2}, \frac{W}{2}\right]$. W is thus the width of the distribution and controls the disorder strength. This changes the Hamiltonian operator, but not any of the $h^{\alpha_1\cdots}$. It has the general effect of randomly shifting the energy levels around, broadening the sharp features of the conductivity. This property can be clearly seen in the first example (see fig. 3), where we study the effect of varying strengths of disorder in the photogalvanic effect [11] for GG. A direct consequence of this broadening is the emergence of an optical response in regions where it used to be zero.

4.2 Vacancies

As our second example of disorder, we employ vacancies, a more realistic model that attempts to reproduce natural defects in the lattice. At random with a probability c,

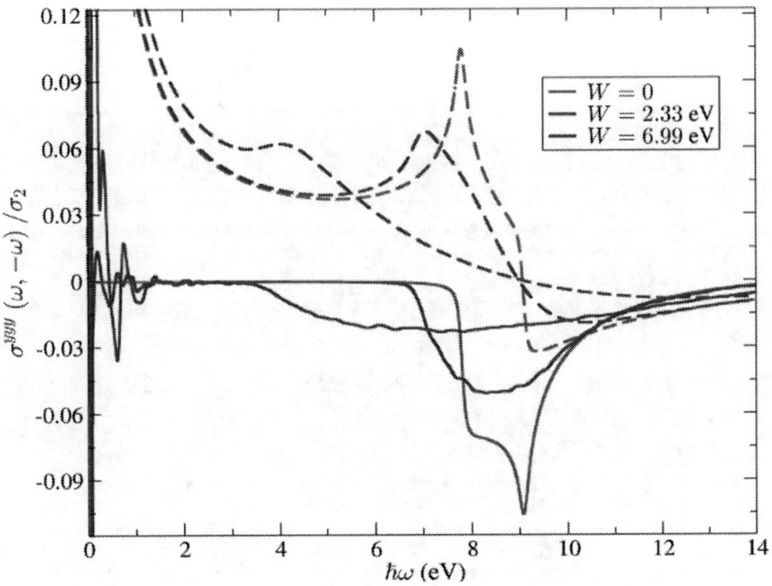

Figure 3. Photogalvanic effect for gapped Graphene in the presence of Anderson disorder of varying strength W and a broadening parameter of $\gamma = 23$ meV. The gap is $\Delta = 7.8$ eV, the Fermi energy is $\varepsilon_F = 0$ eV and the lattice size is $L = 2048$. Number of Chebyshev polynomials: $M = 512$. The dashed lines represent the imaginary part of the conductivity.

atoms are removed from their respective lattice sites. Numerically, this is achieved by removing all the bonds that connect to these atoms. This alters both the Hamiltonian and all the $h^{\alpha_1 \cdots}$. In fig. 4, we represent the longitudinal Second Harmonic Generation (SHG) in GG with varying concentration of vacancies. The effect of this kind of disorder is very different from Anderson disorder. The general broadening of the sharp features in the conductivity is still present, but we no longer see any response below the gap, unlike Anderson disorder.

5 Conclusion and future work

A basis-independent formula for the second-order response is an invaluable tool for the study of realistic systems. It opens up the possibility of exploring the effect of disorder in a fully non-perturbative way. A logical next step from this work is the implementation and study of more elaborate types of disorder, such as adatoms and structural defects. Another interesting application is the study of the second-order response of systems subject to magnetic fields.

6 Acknowledgments

The authors acknowledge financing of Fundação da Ciência e Tecnologia, of COMPETE 2020 program in FEDER component (European Union), through projects POCI-01-0145-FEDER-028887 and UID/FIS/04650/2013. The author S. M. J. also acknowledges financing through PhD grant PD/BD/142798/2018.

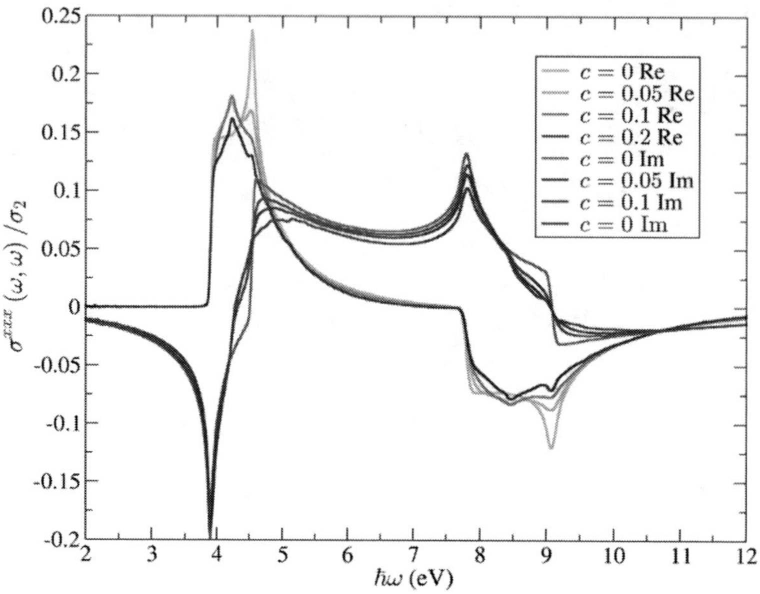

Figure 4. Second-harmonic generation in gapped Graphene for a varying concentration of vacancies and $\gamma = 2.3\text{meV}$. The blue (red) curves represent the real (imaginary) part of the conductivity. The darker curves have a higher concentration of vacancies. All the parameters are the same as in Figure 3.

References

[1] S.M. João, M. Anđelković, L. Covaci, T.G. Rappoport, J.M.V.P. Lopes, A. Ferreira, Royal Society Open Science **7**, 191809 (2020)

[2] C. Aversa, J. Sipe, Physical Review B **52**, 14636 (1995)

[3] J.E. Sipe, E. Ghahramani, Phys. Rev. B **48**, 11705 (1993)

[4] J. Sipe, A. Shkrebtii, Physical Review B **61**, 5337 (2000)

[5] J.L. Cheng, J. E. Sipe, S. W. Wu, C. Guo, APL Photonics **4**, 034201 (2019)

[6] G.B. Ventura, D.J. Passos, J.M.B. Lopes dos Santos, J.M. Viana Parente Lopes, N.M.R. Peres, Phys. Rev. B **96**, 035431 (2017)

[7] D.J. Passos, G.B. Ventura, J.M.V.P. Lopes, J.M.B.L. dos Santos, N.M.R. Peres, Physical Review B **97** (2018), 1712.04924

[8] S.M. João, J.M. Viana Parente Lopes, Journal of Physics: Condensed Matter **32**, 125901 (2020)

[9] A. Weisse, G. Wellein, A. Alvermann, H. Fehske, Reviews of modern physics **78**, 275 (2006)

[10] L.V. Keldysh, Zh. Eksp. Teor. Fiz. **47**, 1018 (1964)

[11] M. Glazov, S. Ganichev, Physics Reports **535**, 101 (2014), high frequency electric field induced nonlinear effects in graphene

EPJ Web of Conferences **233**, 03003 (2020) https://doi.org/10.1051/epjconf/202023303003
CMPNC 2019

Multi-orbital physics of edge-magnetism in a Hubbard model of transition-metal dichalcogenide nanoribbons

Comparing Mean Field Theory and Determinant Quantum Monte Carlo

Francisco Brito[1,*], *Eduardo* V. Castro[1,**], and *João* M. V. P. Lopes[1,***]

[1]Centro de Física das Universidades do Minho e Porto, Departamento de Física e Astronomia, Faculdade de Ciências, Universidade do Porto, 4169-007 Porto, Portugal

Abstract. We study the emergence of edge-magnetism in zigzag nanoribbons from electron-electron interactions in a minimal 3-band tight-binding description of transition-metal dichalcogenide monolayers with an intra-orbital Hubbard term. We explain the influence of each orbital in the magnetic ordering, comparing the results of Mean Field Theory and determinant Quantum Monte Carlo. We focus on a gapped edge-antiferromagnetic phase appearing at three-quarter edge-filling for realistic values of the interaction.

1 Introduction

Transition-metal dichalcogenides (TMDs) are part of the landscape of 2D materials beyond graphene [1]. These materials have a number of possible technological applications [2–4]. They are advantageous mainly because, unlike graphene, they are semiconductors in the monolayer form, with an intrinsic gap that exceeds 1 eV [5, 6]. Additionaly, TMDs are good candidates for use in spin and valleytronics, where it is crucial to control the electronic spin and the valley degree of freedom.

We focus on molybdenum disulfide, MoS_2, a part of the family of group 6 TMDs, denoted MX_2, with $M = Mo, W$ and $X = S, Se, Te$. We consider an atomically thin (~ 6.5 Å), trigonal prismatic (2H) structure with ABA stacking. The structure consists of a layer of Mo-atoms in a triangular lattice, sandwiched between two layers of S-atoms, also organized in triangular lattices, as shown in the a) and b) panels of figure 1. We assume that the layers are much longer on one direction than on the other. This type of geometry is known as a nanoribbon since the length along the narrow direction is typically of the order of the nanometer.

The typically strong electron-electron interactions in transition-metal atoms potentially lead to the emergence of magnetism in TMDs. In fact, a ferromagnetic spin-valley polarized phase was recently predicted in a minimal model of a hole-doped TMD monolayer [7]. In this work, we probe TMD nanoribbons for magnetic order emerging from intra-orbital Hubbard interactions. Using Mean Field Theory (MFT) and determinant Quantum Monte Carlo (detQMC), we predict and characterize an edge-antiferromagnetic phase, appearing in the

*e-mail: francisco.brito@fc.up.pt
**e-mail: evcastro@fc.up.pt
***e-mail: jlopes@fc.up.pt

minimal model used here when the edges are at three-quarter filling, that is when 3/4 of the electronic edge-states are occupied.

2 Model

The model we use is based on the 3-band tight-binding (3BTB) model introduced in [8] with an additional Hubbard term that penalizes double orbital occupation. The Hamiltonian reads

$$\mathcal{H} = \mathcal{H}_K + \mathcal{H}_V = \sum_{\substack{i,j \\ \alpha,\beta,\sigma}} c^\dagger_{i,\alpha,\sigma} K^\sigma_{\alpha\beta}(\mathbf{R}_i - \mathbf{R}_j) c_{j,\beta,\sigma} + U \sum_{i,\alpha} n_{i,\alpha,\uparrow} n_{i,\alpha,\downarrow}, \tag{1}$$

where i, j are sites on the triangular lattice, σ is the electron spin, and α, β are orbitals from the basis set $\{d_{z^2}, d_{xy}, d_{x^2-y^2}\}$. $K_{\alpha\beta}(\mathbf{R}_i - \mathbf{R}_j) = \langle \phi_\alpha(\mathbf{R}_i)|\mathcal{H}|\phi_\beta(\mathbf{R}_j)\rangle$ are the hopping integrals (which depend on the orbitals, spin, and which of the 6 neighbors the electron hops to, with on-site terms also being allowed) [8]. Here, $\mathbf{R}_{i,j}$ are lattice vectors and $|\phi_{\alpha,\beta}(\mathbf{R}_{i,j})\rangle$ are Wannier states corresponding to orbitals α, β localized at $\mathbf{R}_{i,j}$. U is the Hubbard interaction. The 3BTB model captures edge-physics well. It is based on the symmetries of the monolayer and the fact that at low energies, both band edges have main contributions from the d_{z^2},d_{xy}, and $d_{x^2-y^2}$ orbitals of the M-atoms. Spin-orbit coupling (SOC) leads to a splitting of the energy bands, with $\varepsilon_{k,\sigma} = \varepsilon_{-k,-\sigma}$ (as imposed by time-reversal symmetry and shown in the panel d) of figure 1). However, this energy scale is much smaller than the typical Hubbard interaction required for magnetic ordering. Thus, we neglect SOC in what follows.

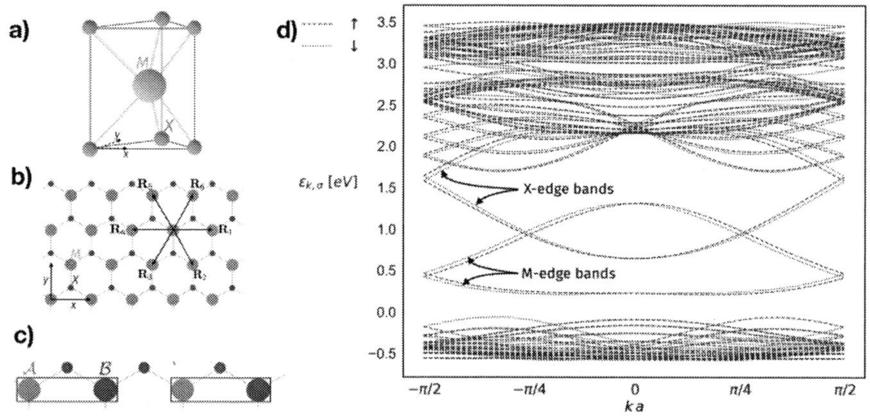

Figure 1. a) Trigonal prismatic (2H) phase of a TMD monolayer. b) Projection onto the xy plane, yielding a honeycomb lattice. Neighboring M-atoms are labeled by lattice vectors $\mathbf{R}_{i=1,2,\ldots,6}$. c) Grouping Mo-atoms two by two on each row of the ribbon, we define two sublattices $\mathcal{A}$ (orange) and $\mathcal{B}$ (red). d) Band structure of the (non-interacting) 3BTB model for a ribbon of 10 Mo-atoms of width, with spin-orbit coupling.

In MFT, the Hamiltonian of equation (1) is replaced by the Hamiltonian that best describes the physics of the system, while being quadratic in the operators $c^{(\dagger)}_{i,\alpha,\sigma}$, $\mathcal{H}_{\mathrm{MF}}$. We minimize the variational grand potential Ω_V, an upper bound on the real grand potential Ω:

$$\Omega \leq \langle \mathcal{H} \rangle_{MF} - TS_{MF} - \mu \langle N \rangle_{MF} \equiv \Omega_V, \tag{2}$$

where T is the temperature of the system, S_{MF} and $\langle \mathcal{H} \rangle_{MF}$ are, respectively, the entropy and energy computed with the mean field Hamiltonian, i.e. using the distribution $\rho_{MF} = e^{-\beta \mathcal{H}_{MF}}$, and μ is the chemical potential. The total particle number operator $\mathcal{N}$ commutes with the Hamiltonian, $[\mathcal{H}, \mathcal{N}] = 0$. Applying the variational principle to equation (1) and assuming spin is still a good quantum number, we obtain

$$\mathcal{H}_{\mathrm{MF}} = \mathcal{H}_K + U \sum_{i,\alpha,\sigma} n_{i,\alpha,\sigma} \langle n_{i,\alpha,-\sigma} \rangle \tag{3}$$

3 Methods

Finding a self-consistent solution $\langle n_{i,\alpha,\sigma} \rangle$ corresponds to finding a minimum of Ω_V. The global minimum corresponds to the mean field solution. We wish to reach the mean field solution starting with unbiased random or paramagnetic initial mean fields. To do so, we successively diagonalize the mean field Hamiltonian and use the obtained spectrum and eigenfunctions to recompute the mean fields, repeating the process until the mean fields vary only negligibly from iteration to iteration [9]. We use the grand canonical ensemble, fixing μ and obtaining the edge filling, which we fix using a bissection method. Starting from 8 random and 8 paramagnetic uncorrelated initial conditions, we find self-consistent solutions and choose the one that reaches the lowest variational grand potential. To improve convergence,

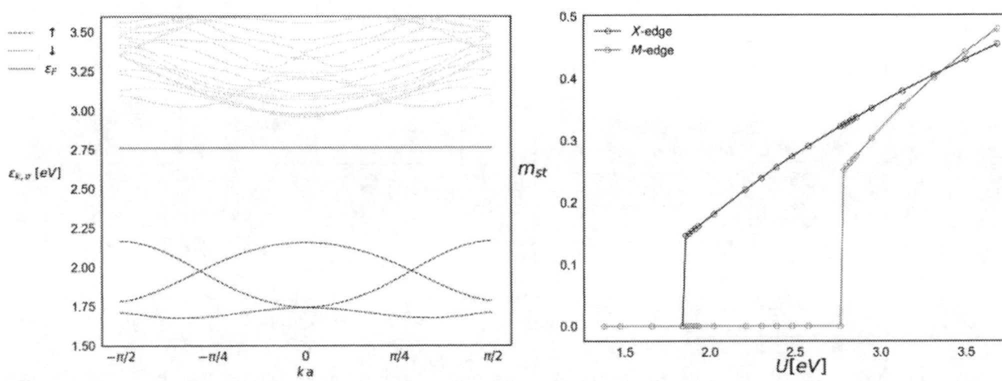

Figure 2. Left: Part of the mean field band structure for $U = 2.94\,eV$ at zero temperature for an infinite system along the longitudinal direction, with 10-Mo atoms along the transverse direction, ignoring spin-orbit coupling. Note that six of the eight spin-degenerate bands corresponding to the Mo- and S-terminated edges are occupied, hence the system is at three-quarter edge filling. Unoccupied bands are faded. Right: Staggered magnetization at the nonequivalent M- and X-edges of the same MoS_2 nanoribbon as a function of the Hubbard interaction U, obtained with MFT at zero temperature.

we use annealing, starting from a high temperature and decreasing it smoothly until the desired temperature is reached. The dimensionality of our minimization problem is reduced by using the longitudinal translation symmetry of the ribbon, while considering 2 sublattices to allow antiferromagnetic solutions (see panel c) of figure 1).

The determinant Quantum Monte Carlo method is based on a mapping to a higher-dimensional non-interacting system using an auxiliary field. The Hubbard interaction is mediated by this discrete, and binary field **h**. The mapping leads to a problem of independent fermions coupled to **h**, and the fermionic part of the partition function may be traced out explicitly [10, 11]:

$$Z = \mathrm{Tr}[e^{-\beta\mathcal{H}}] \approx \mathrm{Tr}[\prod_{l=0}^{L-1} e^{-\Delta\tau\mathcal{H}_l}] = \sum_{\{\mathbf{h}\}}\sum_{\text{fermion}} e^{-S} = \sum_{\{\mathbf{h}\}} p(\{\mathbf{h}\}) \propto \mathrm{Tr}_{\mathbf{h}}\Big[\prod_{\sigma} det[\mathbf{I}+\prod_{l=L-1}^{0}\mathbf{B}_{l,\sigma}(\widetilde{\mathbf{h}}_l)]\Big], \tag{4}$$

where S is a fermion-field action, $p(\{\mathbf{h}\})$ is the statistical weight of a configuration of the field, $\mathbf{I}$ is the identity matrix, imaginary-time is discretized in $L = \beta\Delta\tau$ slices labeled by index l, and $\mathbf{B}_{l,\sigma}(\widetilde{\mathbf{h}}_l)$ are square matrices with the dimension of the size of the system that depend on the parameters of the model and on the field $\mathbf{h}$ at each imaginary-time slice l. We use the grand canonical ensemble: $\mathcal{H}_K \to \mathcal{H}_K - \mu\mathcal{N}$. Using Wick's theorem, observables may be written in terms of the Green's matrices for a fixed configuration of the field $\mathbf{h}$, $\mathbf{G}^{\sigma}(\mathbf{h}) = [\mathbf{I} + \prod_{l=L-1}^{0}\mathbf{B}_{l,\sigma}(\widetilde{\mathbf{h}}_l)]^{-1}$. The spin-spin correlator is measured by sampling configurations of $\mathbf{h}$ using Monte Carlo and then averaging over the field to obtain an estimate of $\left\langle S^z_{i,\alpha}S^z_{j\beta}\right\rangle \equiv \left\langle (n_{i,\alpha,\uparrow} - n_{i,\alpha,\downarrow})(n_{j\beta,\uparrow} - n_{j\beta,\downarrow})\right\rangle$.

4 Results

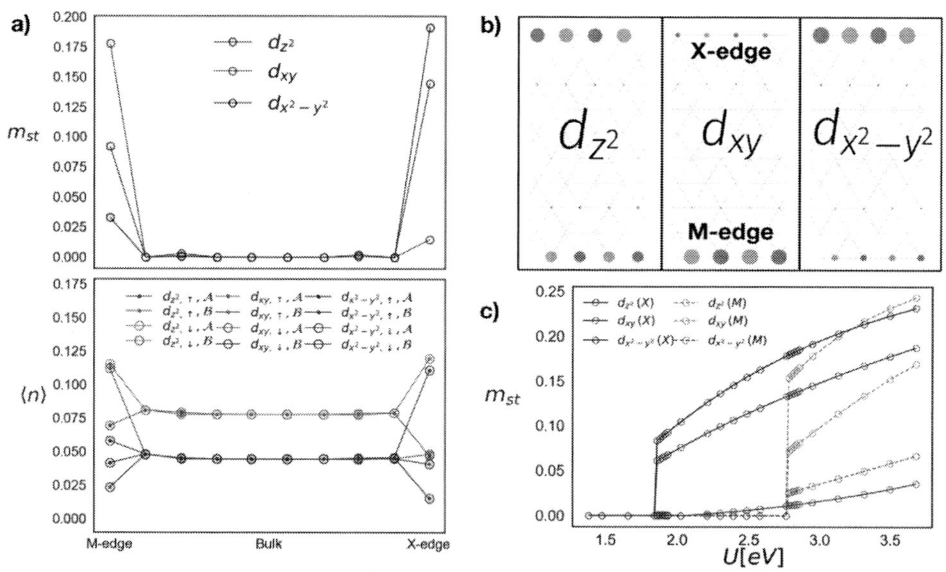

Figure 3. a) Staggered magnetization (top) and orbital- and spin-resolved electron density (bottom) for each row of a ribbon with 10 Mo-atoms of width, with $U = 2.94eV$ at $T = 0$. b) Antiferromagnetic order in MFT at 3/4 edge filling for the same parameters. The size of the circles indicates the magnitude of the spins. Red means spin-up and blue spin-down. c) Orbital-resolved staggered magnetization at the Mo-(M) and S(X)-terminated edges as a function of U for the same ribbon.

Here, we compare the results obtained with MFT and detQMC at three-quarter edge filling for a MoS_2 nanoribbon, focusing on the role played by each orbital in the emergence of edge-antiferromagnetic order. In the left panel of figure 2, we show part of the mean field band structure, and we show how the staggered magnetization depends on U at $T = 0$ (see right panel of figure 2). Although this is not shown, in MFT, we have first order phase transitions, occurring, for the most part, above or near room temperature. The transitions to antiferromagnetic phases occur for different critical interactions on the two edges: $U_c^X \approx 1.84\,eV$,

$U_c^M \approx 2.76\,eV$. The X-edge is more susceptible to antiferromagnetic order since the transition occurs sooner: $U_c^X < U_c^M$.

Different orbitals are responsible for the magnetism of the two nonequivalent edges. In panels a) (bottom) and b) of figure 3, we see that the imbalance in the number of electrons occupying the $d_{x^2-y^2}$ orbital is the main cause of magnetic order on the X-edge, and similarly for the d_{xy} orbital for order on the M-edge. The d_{z^2} orbital contributes significantly for both edges, but more on the X-edge. Thus, the staggered magnetization is higher on the X-edge. This could justify the fact that $U_c^X < U_c^M$ (see panel c) of figure 3 from an orbital perspective). From the point of view of the mean-field bands, it is obvious that the AF ordering of the X-edge implies the opening of a gap at the Fermi level and, consequently the minimization of the variational potential of the system. As can be seen in panel a) of figure 3 (bottom), the orbitals have equal occupations on the bulk, inhibiting magnetic order. However, at the edges, sublattices $\mathcal{A}$ and $\mathcal{B}$ have different occupations, leading to antiferromagnetic order.

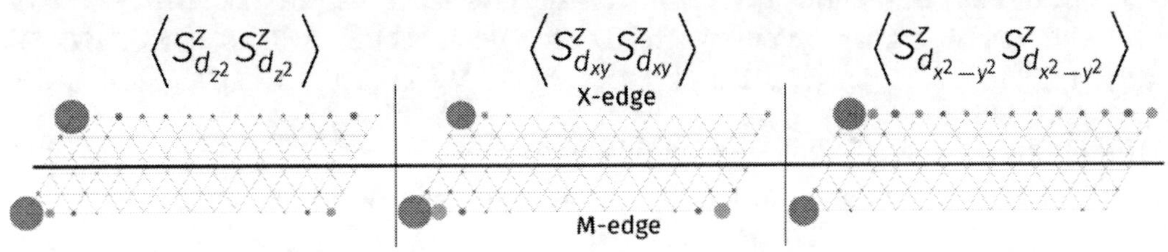

Figure 4. Orbital-resolved spin-spin correlations from detQMC for a 14×5 MoS_2 nanoribbon (only 3 rows are shown, since the correlations decay quickly) for $U = 2.94eV$ and a temperature of $T \approx 267K$. The size of the circles indicates the magnitude of the correlation. Red means a positive correlation and blue a negative one. In the top panel, we show the correlations with respect to the leftmost site of the X-edge and in the down panel, the correlations with respect to the leftmost site of the M-edge.

Our detQMC results confirm the predictions of MFT. In figure 4, we can see that while antiferromagnetic spin-spin correlations decay in detQMC, unlike in MFT (in which case they are constant), they decay slower for the d_{z^2} and $d_{x^2-y^2}$ orbitals at the X-edge, and for the d_{z^2} and d_{xy} orbitals at the M-edge, decaying quickly in the bulk in both cases.

5 Conclusion

We studied the emergence of edge-antiferromagnetism from the Hubbard interaction in zigzag TMD nanoribbons. Using Mean Field Theory and determinant Quantum Monte Carlo, we focused on the multi-orbital aspects of the magnetic ordering, explaining how antiferromagnetism sets in on the transition-metal and chalcogen-terminated edges at three-quarter edge filling, comparing the results obtained with the two methods. We found a greater tendency towards magnetic ordering on the X-edge compared with the M-edge, with a lower critical interaction $U_c^X < U_c^M$ required for order to set in at $T = 0$.

6 Acknowledgements

The authors acknowledge financing from Fundação da Ciência e Tecnologia and COMPETE 2020 program in FEDER component (European Union), through the projects POCI-01-0145-FEDER-02887 and UID/FIS/04650/2013.

References

[1] S. Manzeli, D. Ovchinnikov, D. Pasquier, O.V. Yazyev, A. Kis, Nature Reviews Materials **2**, 17033 (2017)

[2] B. Radisavljevic, A. Radenovic, J. Brivio, V. Giacometti, A. Kis, Nature Nanotechnology **6**, 147 (2011)

[3] F.H.L. Koppens, T. Mueller, P. Avouris, A.C. Ferrari, M.S. Vitiello, M. Polini, Nature Nanotechnology **9**, 780 (2014)

[4] K.F. Mak, J. Shan, Nature Photonics **10**, 216 (2016)

[5] Y. Zhang, T.T. Tang, C. Girit, Z. Hao, M.C. Martin, A. Zettl, M.F. Crommie, Y.R. Shen, F. Wang, Nature **459**, 820 (2009)

[6] K.F. Mak, C. Lee, J. Hone, J. Shan, T.F. Heinz, Physical Review Letters **105**, 136805 (2010)

[7] J.E.H. Braz, B. Amorim, E.V. Castro, Physical Review B **98**, 161406 (2018)

[8] G.B. Liu, W.Y. Shan, Y. Yao, W. Yao, D. Xiao, Physical Review B **88**, 085433 (2013)

[9] H. Feldner, Z.Y. Meng, A. Honecker, D. Cabra, S. Wessel, F.F. Assaad, Physical Review B **81**, 115416 (2010)

[10] J.E. Hirsch, Physical Review B **28**, 4059 (1983)

[11] W. Hanke, Y.V. Kopaev, W. John, Crystal Research and Technology **28**, 28 (1993)

EPJ Web of Conferences **233**, 03004 (2020)
CMPNC 2019
https://doi.org/10.1051/epjconf/202023303004

Electronic Properties of Twisted Bilayer Graphene in the Presence of a Magnetic Field

M. F. C. Martins Quintela[1,*], *J. C. C.* Guerra[1,**], and *S. M.* João[1,***]

[1]Centro de Física das Universidades do Minho e do Porto, CF-UM-UP Departmento de Física e Astronomia, Universidade do Porto, Rua do Campo Alegre 4169-007, Porto, Portugal

Abstract. In AA-stacked twisted bilayer graphene, the lower energy bands become completely flat when the twist angle passes through certain specific values: the so-called "magic angles". The Dirac peak appears at zero energy due to the flattening of these bands when the twist angle is sufficiently small [1-3]. When a constant perpendicular magnetic field is applied, Landau levels start appearing as expected [5]. We used the Kernel Polynomial Method (KPM) [6] as implemented in KITE [7] to study the optical and electronic properties of these systems. The aim of this work is to analyze how the features of these quantities change with the twist angle in the presence of an uniform magnetic field.

1 Introduction

Until very recently, the origin of the magic angles, which are characterized by the extreme flattening of the lowest energy bands, in twisted bilayer graphene (TBG) was unknown. In 2007 and 2012, J. Lopes dos Santos, N. Peres *et al* published two articles [1, 2] where they developed a continuum model for this material. With this work, they were able to find a simple relation between supperlattice vectors with which they could explain the physics of all types of commensurate structures. Very recently, in 2019, A. Vishwanath went over the origin for these magic angles and their periodicity in $\alpha \sim 1/\theta$ [3].

Using KITE, which employs a Chebyshev polynomial-approach to approximate Green's functions, it is possible to calculate the density of states for a multitude of 2D and 3D structures. This allows us to calculate the density of states for several commensurable angles, as well as the optical conductivity and the local density of states [7].

1.1 Moiré Superlattice and Hamiltonian

While the structure of graphene, the honeycomb lattice, is reasonably simple, it becomes quite complex when two graphene sheets are stacked and a relative twist is applied to one of them. For some angles (denominated commensurate angles), a periodic structure emerges with a periodicity much larger than that of the individual graphene lattices. This is the Moiré superlattice, which is visible in Figure 1 with central AA-stacking and a twist angle of 21.798 degrees.

*e-mail: mfcmquintela@fc.up.pt
**e-mail: cguerrajoao@gmail.com
***e-mail: up201206827@fc.up.pt

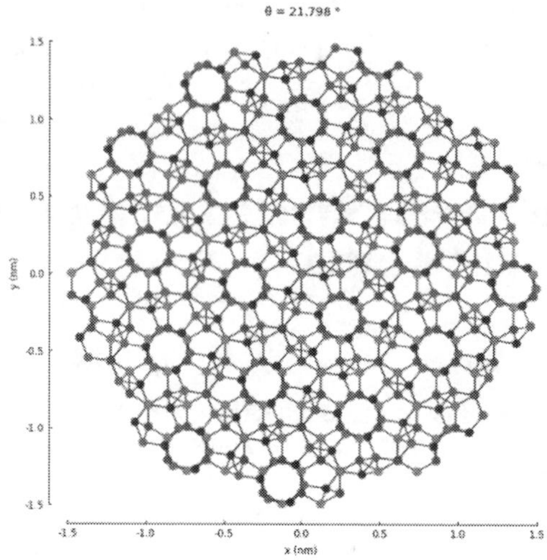

Figure 1. Moiré superlattice for a twist angle of 21.798 degrees

The second quantization tight-binding Hamiltonian [4] can be written as a sum of an intra-layer term and an inter-layer one

$$\begin{aligned} H &= H_{\parallel} + H_{\perp} \\ &= -\sum_{i \neq j, m} \gamma_{ij}^{mm} \left(c_{m,i}^{\dagger} c_{m,j} + \text{h.c.} \right) - \sum_{i,j,m} \gamma_{ij}^{m,m+1} \left(c_{m,i}^{\dagger} c_{m+1,j} + \text{h.c.} \right) \end{aligned} \tag{1}$$

where $c_{m,i}^{\dagger}$, $c_{m,i}$ are, respectively, the creation and annihilation operators of a p_z state in the atomic site i in layer m, and γ_{ij}^{mm} is the in-plane hopping integral between these atomic sites. As we are only considering a nearest neighbour intra-layer hopping, γ_{ij}^{mm} will be zero when the atomic sites are not first-neighbours, and, when they are, will be given by $V_{pp\pi}^{0}$ (the nearest neighbours hopping parameter for single-layer graphene, obtained via DFT). For the inter-layer term, we are considering the hopping integral to be of the form

$$t(r) = V_{pp\sigma}^{0} \left(1 - \frac{c_0^2}{r^2} \right) e^{-\frac{r - c_0}{\delta}} \tag{2}$$

where r is the 3D distance between the two atomic sites, c_0 is the inter-layer distance , $V_{pp\sigma}^{0}$ is the hopping integral between two vertically-stacked atoms in untwisted bilayer graphene, and δ is a parameter that modulates the hopping cutoff at long distances (also obtained via DFT).

2 Twisted Bilayer Graphene in a Magnetic Field

Due to the effect of a perpendicular magnetic field in monolayer graphene [5], its energy spectrum becomes quantized into Landau Levels, whose expression is given by

$$\epsilon_n = \text{sign}(n) \frac{\hbar v_F \sqrt{2}}{l_B} \sqrt{|n|} \tag{3}$$

where v_F is the Fermi velocity, and the magnetic length l_B is defined as

$$l_B = \sqrt{\frac{\hbar}{eB}}. \tag{4}$$

This also creates specific selection rules for the transitions between energy levels, which will modify its optical properties.

2.1 Density of States and Landau Levels

To study the dependence on the twist angle of the density of states, we first calculated it using KITE [7] for several commensurate angles without any magnetic field. In agreement with [8], the Dirac Peak, which is caused by the forming of a flat-band at zero energy, appears for sufficiently low twist angles (less than 3 degrees), around which two gaps open up. This is visible in Figure 2.

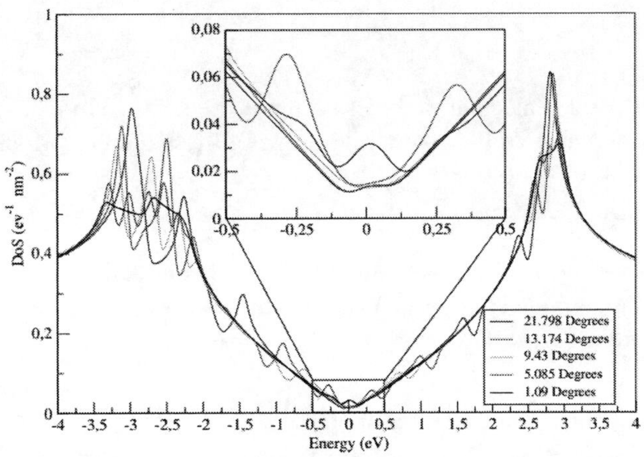

Figure 2. Density of states for several twist angles

When the magnetic field is added using the Peierls substitution [9]

$$t\left(r_j - r_i\right) \rightarrow \tilde{t}\left(r_j - r_i\right) = t\left(r_j - r_i\right) e^{-i\frac{e}{\hbar}\int_{r_i}^{r_j} \mathbf{A}(r')\cdot d\mathbf{r}'} \tag{5}$$

the energy levels change qualitatively. The crossing of each Landau Level creates well-known peaks in the density of states [11]. To analyze the differences, the densities of states for both $B = 0\,\mathrm{T}$ and for $B = 24\,\mathrm{T}$ are plotted together for four different twist angles.

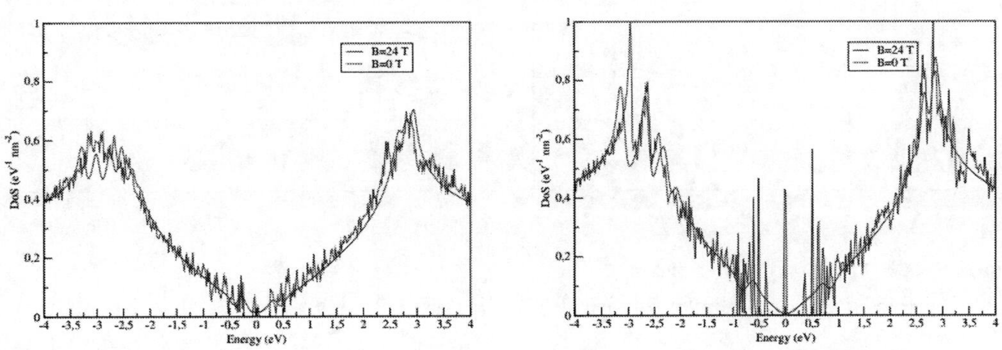

Figure 3. Comparison of the densities of states for a twist angle of 5.085 and 9.430 degrees

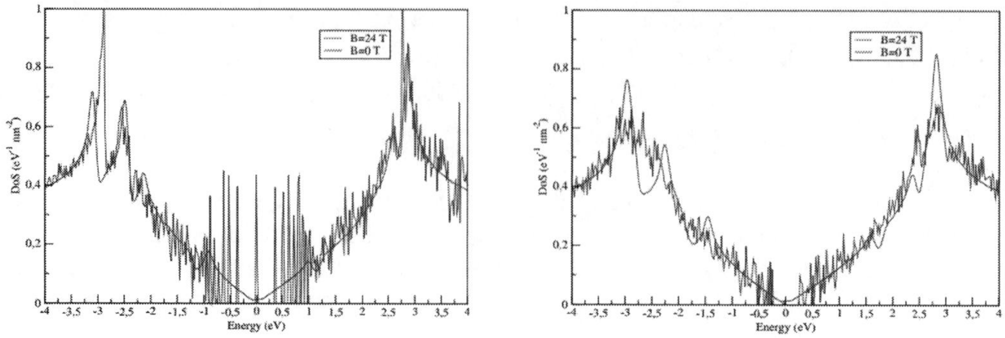

Figure 4. Comparison of the densities of states for a twist angle of 13.174 and 21.798 degrees

The above-mentioned peaks due to the crossing of the LLs are more clearly visible for 9.43 and 13.174 degrees. Because of this, we will use one of these angles for our study.

2.2 DC Conductivity

To finalize this proceeding, we will take a look at the DC conductivity in both the xx and the xy directions, which were obtained through KITE. For this purpose, KITE employs the Kubo-Bastin formula in the velocity gauge:

$$\sigma_{\mathrm{DC}}^{\alpha\beta} = \frac{-e^2}{V}\int_{-\infty}^{\infty} d\epsilon f(\epsilon)\,\mathrm{Tr}\left[\delta(\epsilon-\hat{H})\hat{v}^{\alpha}\frac{dG^{+}}{d\epsilon}(\epsilon)\hat{v}^{\beta} - \delta(\epsilon-\hat{H})\hat{v}^{\beta}\frac{dG^{-}}{d\epsilon}(\epsilon)\hat{v}^{\alpha}\right] \tag{6}$$

where $G^{\pm}$ are the retarded/advanced Green's functions respectively, defined by

$$G^{\pm}(\epsilon) = \frac{1}{\epsilon - \hat{H} \pm i0^{+}}. \tag{7}$$

By construction of this formula, the imaginary part of the DC-conductivity will be zero. As such, we will focus only on the real part of both $\sigma_{\mathrm{DC}}^{xx}$ and $\sigma_{\mathrm{DC}}^{xy}$, which we plot for a twist angle of 9.43 degrees and a magnetic field of 24 T in Figure 5.

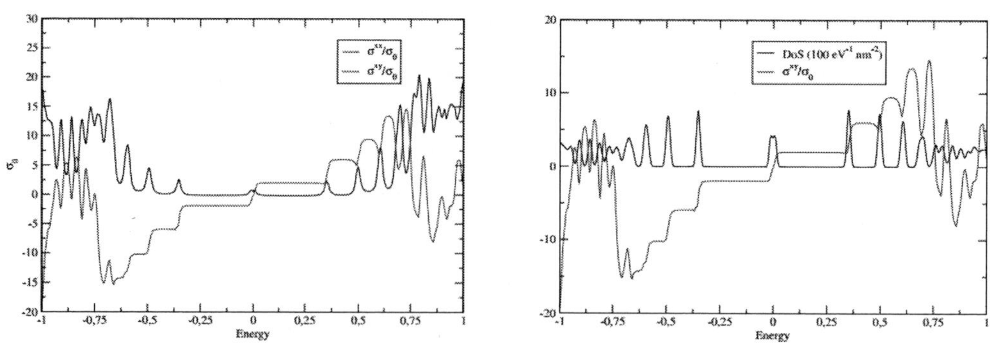

Figure 5. (Left) Real part of $\sigma_{\mathrm{DC}}^{xx}$ and $\sigma_{\mathrm{DC}}^{xy}$
(Right) Real part of $\sigma_{\mathrm{DC}}^{xy}$ juxtaposed onto the (rescaled) Density of States

The qualitative behaviour of these quantities is as expected, with their peaks and plateaus upon the crossing of each Landau Level (visible by the sudden spike in the density of states).

3 Conclusions and Future Work

Twisted bilayer graphene is a very promising material, with a vast array of useful properties, amongst which is the flattening of the energy bands due to the vanishing of the Fermi velocity, which leads to superconducting states [11]. We were able to obtain the DC conductivity in both the xx and the xy directions in the presence of a magnetic field through KITE. In the future, we would want to study the angle dependence of these quantities, as was done previously for both the presence of disorder and/or a perpendicular electric field [12]. One point that will be of interest for future study is doing this same comparison for optical conductivity [13], both first and second order.

4 Acknowledgements

The authors acknowledge financing of Fundação da Ciência e Tecnologia, of COM-PETE 2020 program in FEDER component (European Union), through projects POCI-01-0145-FEDER-028887 and UID/FIS/04650/2013. The authors also acknowledge financial support from Fundação para a Ciência e Tecnologia, Portugal, through national funds, co-financed by COMPETE-FEDER (grant M-ERA- NET2/0002/2016 – UltraGraf) under the Partnership Agreement PT2020.

FCT
Fundação para a Ciência e a Tecnologia
MINISTÉRIO DA CIÊNCIA, TECNOLOGIA E ENSINO SUPERIOR

References

[1] Lopes dos Santos, J., Peres, N., Castro Neto, A., Physical Review B 86, 155449 (2012)
[2] Lopes dos Santos, J., Peres, N., Castro Neto, A., Physical Review Letters 99, 256802 (2007)
[3] Vishwanath, A., Kruchkov, A., Tarnopolsky, G., Physical Review Letters 122, 106405 (2019)
[4] Moon, P., Koshino, M., Phys. Rev. B 85, 195458 (2012)
[5] Goerbig, M. O., Reviews of Modern Physics 83, 1193 (2011)
[6] Weiße, A., Wellein, G., Alvermann, A., Reviews of Modern Physics 78, 275 (2016)
[7] Lopes, J.V., João, S.M., Ferreira, A., Covaci, L., Andelkovic, M., Rappoport. T., (2018) Quantum KITE. Website: https://quantum-kite.com/
[8] Moon, P., Koshino, M., Physical Review B 85, 195458 (2012)
[9] Peierls, R., Z. Phys. 80: 763–791 (1933)
[10] Anh Le, H., Nam Do, V., Physical Review B 97, 125136 (2018)
[11] Cao, Y., Fatemi, V., Fang, S., et al, Nature 556, 43 (2018)
[12] Anđelković, M., Covaci, L., Peeters, F. M., Phys. Rev. Materials 2, 034004 (2018)
[13] Catarina, G., Amorim, B., Castro, E. V., Lopes, J. V., Peres, N., Handbook of Graphene: Volume 3, Chap. 6, pp. 177-230 (2019)

Energy loss by fast-travelling charged particles traversing two-dimensional materials

Jaime E. Santos[1,*], *Mikhail* Vasilevskiy[1,2], *Nuno* M. R. Peres[1,2], and *Antti-Pekka* Jauho[3]

[1]Centro de Física e Departamento de Física, Universidade do Minho, 4710-057 Braga, Portugal
[2]International Iberian Nanotechnology Laboratory, Avenida Mestre José Veiga s/n, 4715-330 Braga, Portugal
[3]Center for Nanostructured Graphene (CNG), Department of Physics DTU, building 345C, 2800 Kgs. Lyngby, Denmark

Abstract. We consider the problem of the radiation losses by fast-traveling particles traversing two-dimensional (2d) materials or thin films. After reviewing the screening of electromagnetic fields by two dimensional conducting materials, we obtain the energy loss by a fast particle traversing such a material or film. In particular, we discuss the pattern of radiation emitted by monolayer graphene treated within a hydrodynamic approximation. These results are compared with recent published results using similar approximations and, having in mind a potential application to particle detection, we briefly discuss how one can improve on the signals obtained by using other two-dimensional materials.

1 Introduction

The problem of the interaction of the electromagnetic field with a 2d conductor has a long and distinguished history, going back to the 1950s [1–5]. Such problem gained a renewed interest in the context of the transport properties of graphene and related materials [6, 7]. These materials hold the promise of yielding devices with tailored optical properties, among others [8–10].

We briefly review here the formulation of such a problem, as well as the related issue of the calculation of the conductivity of a 2d material, so as to apply the underlying results to the problem of radiation losses by fast-particles traversing graphene. We conclude that the signal emitted by graphene is too small to be of use, but such a result may be modified by considering a 2d material with a much larger in-plane AC conductivity.

2 Definition of the screening tensor

We consider a 2d material, e.g. graphene, embedded in a dielectric medium of relative electric and magnetic permittivities, ε_m and μ_m. The said medium occupies the $z = 0$ plane. The simplest case is that of a suspended material in the vacuum, when $\varepsilon_m = \mu_m = 1$. If a set of external sources generate an applied electric field $\mathbf{E}^{\text{ext}}(\mathbf{r}, t)$, surface densities of charge and

*e-mail: jaime.santos@fisica.uminho.pt

current are induced in the material. At the linear response level, the density of current is given by

$$J_c(\mathbf{q},\omega) = \sigma_L(\mathbf{q},\omega)\mathbf{E}_{\parallel L}(z=0,\mathbf{q},\omega) + \sigma_T(\mathbf{q},\omega)\mathbf{E}_{\parallel T}(z=0,\mathbf{q},\omega), \tag{1}$$

where $\sigma_{L,T}(\mathbf{q},\omega)$ are the components of the conductivity tensor of the 2d material and $\mathbf{E}_{\parallel L,T}(z = 0,\mathbf{q},\omega)$ are the corresponding components of the local electric field along the plane of the material (respectively parallel and perpendicular to $\mathbf{q}$). Here, we employ a mixed real-Fourier space representation, which takes advantage of the translation symmetry in the xy plane.

Such field is related to the induced density of current (1) by (with $v = c/\sqrt{\mu_m\varepsilon_m}$)

$$\mathbf{E}_{\parallel}(z=0,\mathbf{q},\omega) = \mathbf{E}^{\text{ext}}_{\parallel}(z=0,\mathbf{q},\omega) + \frac{Z_m}{2\sqrt{1-q^2v^2/\omega^2}}\left[J_c(\mathbf{q},\omega) - \frac{v^2(\mathbf{q}\cdot J_c(\mathbf{q},\omega))\mathbf{q}}{\omega^2}\right], \tag{2}$$

where $\omega > qv$ and where the solution of the Maxwell equations was obtained using the Coulomb gauge. Here, $Z_m = \sqrt{\frac{\mu_0\mu_m}{\varepsilon_0\varepsilon_m}}$ is the impedance of the surrounding medium. As will be explained below, the choice of the gauge is not completely immaterial (for reasons of intelligibility), even if the distribution of the fields is of course independent from it.

Expressing the induced current in terms of the local field using (1), we can obtain a relation between the applied field and the local field. Defining the components of the screening tensor

$$\zeta_{L,T}(\mathbf{q},\omega) = \frac{E^{\text{ext}}_{\parallel L,T}(z=0,\mathbf{q},\omega)}{E_{\parallel L,T}(z=0,\mathbf{q},\omega)}, \tag{3}$$

we obtain the result

$$\zeta_L(\mathbf{q},\omega) = 1 + \frac{1}{2}Z_m\sigma_L(\mathbf{q},\omega)\sqrt{1-q^2v^2/\omega^2}, \tag{4}$$

and

$$\zeta_T(\mathbf{q},\omega) = 1 + \frac{Z_m\,\sigma_T(\mathbf{q},\omega)}{2\sqrt{1-q^2v^2/\omega^2}}, \tag{5}$$

where retardation effects are included.

If $\omega < qv$, one employs the continuation $\sqrt{1-q^2v^2/\omega^2} = i(v/\omega)\sqrt{q^2-\omega^2/v^2}$ in the above formulas, as determined by the condition of causality. As in the case of a 3d medium [11], the knowledge of the conductivity of a 2d material is sufficient to determine its screening properties, even if the relations obtained are different in the former and latter cases.

It can be shown that if a plane electromagnetic wave impinges on the 2d material, the transmission coefficients for its p and s components are related to the components of the screening tensor by $t_{p,s}(\mathbf{q},\omega) = \zeta^{-1}_{L,T}(\mathbf{q},\omega)$ (with $q = \frac{\omega}{v}\sin\theta$, where θ is the angle of incidence of the wave) [12]. Such result can be understood if we interpret the incident wave as the applied field and the transmitted wave (on the other side of the interface) as the local field.

3 Linear response relations and RPA

The relations obtained above are of little use if one does not provide an explicit expression for the components of the conductivity tensor of the 2d material. One describes it by the Hamiltonian of a many-particle system, in which carriers interact via the Coulomb force, to which one adds a term coupling to an external scalar potential and a term coupling to a vector potential, determined self-consistently [13].

$$\widehat{H} = \frac{1}{2m}\sum_i[\widehat{\mathbf{p}}_i - q_e\mathbf{A}_{\parallel}(z=0,\widehat{\rho}_i,t)]^2 + \sum_i U(\widehat{\rho}_i) + \frac{1}{2}\sum_{i\neq j}V(\widehat{\rho}_i - \widehat{\rho}_j) + q_e\sum_i\phi_{\text{ext}}(z=0,\widehat{\rho}_i,t), \tag{6}$$

where q_e is the carriers charge, $U(\rho)$ is the potential of the underlying lattice and $V(\rho) = \frac{q_e^2}{4\pi\epsilon_0\rho}$ is the Coulomb interaction between the charged particles. The potential $\phi_{\rm ext}(z = 0, \rho, t)$ is the applied external potential. At the linear response level in the external scalar potential and the self-consistent vector potential, the induced current and charge densities are given by

$$j_c^\alpha(\mathbf{q},\omega) = q_e^2\left[\left(\chi^{\alpha\beta}(\mathbf{q},\omega) - \frac{n_0}{m}\delta^{\alpha\beta}\right)A_\parallel^\beta(z=0,\mathbf{q},\omega) - \chi^{\alpha 0}(\mathbf{q},\omega)\,\phi_{\rm ext}(z=0,\mathbf{q},\omega)\right], \quad (7)$$

$$\rho_c(\mathbf{q},\omega) = q_e^2\left[\chi^{0\beta}(\mathbf{q},\omega)A_\parallel^\beta(z=0,\mathbf{q},\omega) - \chi(\mathbf{q},\omega)\,\phi_{\rm ext}(z=0,\mathbf{q},\omega)\right], \quad (8)$$

where α, β designate the Cartesian indices (summation over repeated indices is implied) for the components of the current operator and 0 refers to density operator and where these response functions, $\chi^{\alpha\beta}(\mathbf{q},\omega)$, $\chi^{\alpha 0}(\mathbf{q},\omega)$, $\chi^{0\beta}(\mathbf{q},\omega)$ and $\chi(\mathbf{q},\omega)$ can be expressed, in the case of an isotropic system, in terms of only two quantities, the density-density response function, $\chi(\mathbf{q},\omega)$, and the current-current transverse response function, $\chi^T(\mathbf{q},\omega)$.

This approximation scheme, in which the Coulomb interaction is treated exactly, but where the transverse degrees of freedom of the electromagnetic field are treated self-consistently avoids the explicit inclusion of magnetic interactions (of order v^2/c^2) in the Hamiltonian given by equation (6). Furthermore, do note that $A_{\parallel L}^\alpha(z = 0,\mathbf{q},\omega) \neq 0$ in the Coulomb gauge, as $\mathbf{q}$ is merely the projection of the full wave-vector in the xy plane, so we need to keep such a component in equations (7) and (8). This gauge is not only the most adequate to treat the problem in the approximation scheme of Bohm and Pines [13], it also emphasises the distinction between two and three-dimensional materials.

One can show that the conductivities are related to the response functions by

$$\sigma_L(\mathbf{q},\omega) = -\frac{i\omega q_e^2}{q^2}\cdot\frac{\chi(\mathbf{q},\omega)}{1-\frac{q_e^2}{2\epsilon_0\varepsilon_m q}\chi(\mathbf{q},\omega)}, \quad (9)$$

$$\sigma_T(\mathbf{q},\omega) = \frac{q_e^2}{i\omega}\left(\chi^T(\mathbf{q},\omega) - \frac{n_0}{m}\right), \quad (10)$$

where n_0 is the total electronic density in the material.

If one substitutes these relations in equations (4) and (5), one can obtain an expression for the components of the screening tensor in terms of the response functions as well. The zeros of such components determine the eigen-frequencies of the excitations (plasmons) of the 2d material. Static screening may also be considered [14, 15].

These response functions can now be calculated using a given approximation scheme. In the particular case of the Random Phase Approximation (RPA), it is assumed that the system responds to the self-consistent vector and scalar potentials with the response function of the free-carriers. In that case, equations (9) and (10) reduce to

$$\sigma_L(\mathbf{q},\omega) = -\frac{i\omega q_e^2}{q^2}\cdot\chi_0(\mathbf{q},\omega), \quad (11)$$

$$\sigma_T(\mathbf{q},\omega) = \frac{q_e^2}{i\omega}\left(\chi_0^T(\mathbf{q},\omega) - \frac{n_0}{m}\right), \quad (12)$$

where $\chi_0(\mathbf{q},\omega)$ and $\chi_0^T(\mathbf{q},\omega)$ are the response functions of the quasi-particles in the material, in the absence of the Coulomb interaction. One can further read from equation (11) that the irreducible susceptibility, given by the second quotient in equation (9), reduces in the RPA to the response function of free carriers, $\chi_0(\mathbf{q},\omega)$.

An illustrative case is that of neutral graphene, for which one has, in the low wave-vector and low frequency limit [16, 17],

$$\chi_0(\mathbf{q},\omega) = \frac{q^2}{4\hbar v_F\sqrt{q^2-\omega^2/v_F^2}}, \tag{13}$$

$$\chi_0^T(\mathbf{q},\omega) = \frac{n_0}{m} - \frac{v_F}{4\hbar}\sqrt{q^2-\omega^2/v_F^2}, \tag{14}$$

where v_F is the Fermi velocity of the material. The use of equations (11) to (14) in equations (4) and (5) is sufficient to determine the transmission factors of neutral free-standing graphene [12], within the framework of the RPA.

4 Energy loss through radiation by a fast charged particle traversing graphene

For a 2d material excited by an external field, the energy radiated per solid angle and per frequency interval is given by [18]

$$S(\theta,\phi,\omega) = \frac{\mu_0\mu_m\omega^2}{16\pi^3 v}\left(|J_{cL}(\mathbf{q}_\rho,\omega)|^2\cos^2\theta + |J_{cT}(\mathbf{q}_\rho,\omega)|^2\right), \tag{15}$$

where the components of the induced current are given by equation (1), $\mathbf{q}_\rho = \frac{\omega}{v}\sin\theta\,\hat{\rho}$, with $\hat{\rho}$ being the polar versor in the plane, and where θ and ϕ are the altitude and azimuthal angles. Expressing the local electric field in the material, as given in equation (1), in terms of the applied field using the components of the screening tensor, introduced in equations (4) and (5), we obtain

$$S(\theta,\phi,\omega) = \frac{\mu_0\mu_m\omega^2}{16\pi^3 v}\left(\frac{|\sigma_L(\mathbf{q}_\rho,\omega)|^2\,|E^{\text{ext}}_{\parallel L}(z=0,\mathbf{q}_\rho,\omega)|^2}{|\zeta_L(\mathbf{q}_\rho,\omega)|^2}\cos^2\theta + \frac{|\sigma_T(\mathbf{q}_\rho,\omega)|^2\,|E^{\text{ext}}_{\parallel T}(z=0,\mathbf{q}_\rho,\omega)|^2}{|\zeta_T(\mathbf{q}_\rho,\omega)|^2}\right). \tag{16}$$

In order to consider the profile of radiation due to the passage of a fast-travelling charged particle through the 2d material, we substitute the known expressions for the components of the (applied) field produced by the particle, assumed in uniform motion [19], in equation (16). In the case of doped graphene, we may further use in such an equation the expressions for the conductivity obtained in an hydrodynamic approximation [20], which corresponds to the low wave-vector and low frequency limit of the linear response theory discussed in section 3

$$\sigma_L(\mathbf{q},\omega) = \frac{i\omega\sigma_0}{\tau_F\left[\omega^2 - q^2v_F^2/2 + i\omega\left(1/\tau_F + q^2v_F^2\tau_v/4\right)\right]}, \tag{17}$$

$$\sigma_T(\mathbf{q},\omega) = \frac{i\sigma_0}{\tau_F\left[\omega + i\left(1/\tau_F + q^2v_F^2\tau_v/4\right)\right]}, \tag{18}$$

where τ_F and τ_v are quasi-particle lifetimes, $\sigma_0 = \frac{e^2|\varepsilon_F|\tau_F}{\pi\hbar^2}$ is the DC Drude conductivity of graphene, and where ε_F is the Fermi energy of the material (for the case of undoped graphene, the relevant formulas were already given in section 3, within the realm of the RPA). Integrating over frequency, we can obtain the energy emitted per solid angle. A plot of the emission pattern is given in figure 1. Similar results were obtained in [21, 22].

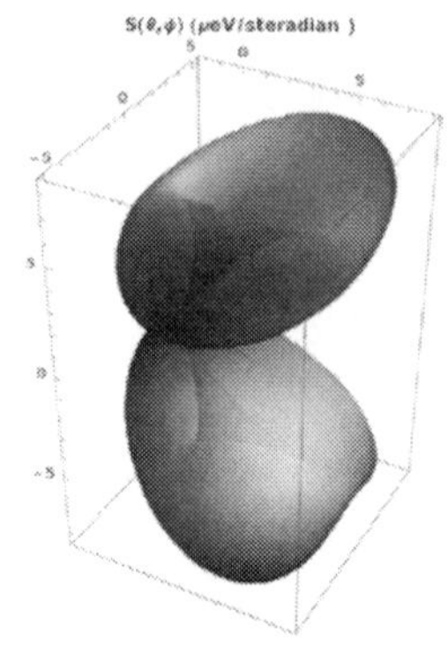

Figure 1. Energy emitted per steradian for an ultra-relativistic electron (with $\beta = 0.9$) traversing a sheet of mono-layer graphene ($\varepsilon_F = 0.5$ eV and $\tau_v = \tau_F = 4.1 \times 10^{-12}$ s) travelling in the xz plane, at an angle of 30° from the vertical.

5 Conclusion

We considered the pattern of radiation emitted by graphene traversed by a fast charged particle. The low values obtained for the emitted energy require, for the purpose of particle detection, the use of materials possessing high effective AC conductivities, possibly 2d semiconductors [23] or thin magnetic films with off-plane magnetisation and a high paramagnetic susceptibility [24].

One should note however, that the emission efficiency of suspended 2d passive materials tends to a limiting value for materials with a high conductivity, as follows from equation (16), since the components of the conductivity tensor appear both in the numerator as well as in the denominator of the two terms of this expression, see equations (4) and (5). Moreover, the real part of the screening factors is always larger than one in the region $\omega > qv$, as the real part of the conductivity is always positive in thermodynamic equilibrium, and hence no resonances are to be seen in this region of frequency space.

Acknowledgements: We acknowledge discussions with K. Thygesen, K. Kasbjerg, S. Mikhailov, B. Amorim, P. Cosme, H. Terças, P. Ribeiro and S. Kirchner. J.E.S. thanks support by the European Structural and Investment Funds in the FEDER component, through the Operational Programme for Competitiveness and Internationalization (COMPETE 2020) [under the Project GNESIS – Graphenest's New Engineered System and its Implementation Solutions; Funding Reference: POCI-01-0247-FEDER-033566], European Regional Development Fund, and would also like to thank the hospitality of the Center for Nanostructured Graphene at the Technical University of Denmark, where most of this research was conducted. The CNG is sponsored by the Danish National Research Foundation through Project No. DNRF103. This work was also supported by the Portuguese Foundation for Science and Technology (FCT) in the framework of the Strategic Funding UID/FIS/04650/2019.

References

[1] R. H. Ritchie, Phys. Rev. **106**, 874 (1957)
[2] F. Stern, Phys. Rev. Lett **18**, 546 (1967)
[3] K. W. Chiu and J. J. Quinn, Phys. Rev. **9**, 4724 (1974)
[4] J. E. Sipe, V. C. Y. So, M. Fukui and G. I. Stegeman, Phys. Rev. **21**, 4389 (1980)

[5] P. Hawrylak, G. Eliasson and J. J. Quinn, Phys. Rev. **37**, 10187 (1988)

[6] S. A. Mikhailov and K. Ziegler, Phys. Rev. Lett. **99**, 016803, (2007)

[7] V. N. Kotov, B. Uchoa, V. M. Pereira, F. Guinea and A. H. Castro Neto, Rev. Mod. Phys. **84**, 1067 (2012)

[8] F. Xia, H. Wang, D. Xiao, M. Dubey and A. Ramasubramaniam, Nature Photonics **8**, 899 (2014)

[9] K. S. Novoselov, A. Mishchenko, A. Carvalho and A. H. Castro Neto, Science **353**, aac9439 (2016)

[10] S. Xiao, X. Zhu, B.-H. Li and N. A. Mortensen, Front. Phys. **11**, 117801 (2016)

[11] D. Pines and P. Noziéres, *The theory of quantum liquids, vol. I* (Walter Benjamin, New York, 1966) 251

[12] T. Stauber, N. M. R. Peres and A. K. Geim, Phys. Rev. B **78**, 085432 (2008)

[13] D. Bohm and D. Pines, Phys. Rev. **82**, 625 (1951)

[14] T. Ando, J. Phys. Soc. Jpn., **75**, 074716 (2006)

[15] P. Cudazzo, I. V. Tokatly and A. Rubio, Phys. Rev. B **84**, 085406 (2011)

[16] K. W.-K. Shung, Phys. Rev. B **34**, 979 (1986)

[17] A. Principi, M. Polini, and G. Vignale, Phys. Rev. B **80**, 075418 (2009)

[18] J. Schwinger, L. L. DeRaad, K. A. Milton, and W. Tsai, *Classical Electrodynamics* (Perseus Books, Reading, Massachusetts, 1998) 378

[19] J. D. Jackson, *Classical Electrodynamics* (John Wiley & Sons, New York, 1999) 560

[20] A. Principi, G. Vignale, M. Carrega, and M. Polini, Phys Rev B **93**, 125410 (2016)

[21] Z. L. Miskovic, S. Segui, J. L. Gervasoni, and N. R. Arista, Phys. Rev. B **94**, 125414 (2016)

[22] K. Akbari, S. Segui, Z. L. Miskovic, J. L. Gervasoni, and N. R. Arista, Phys. Rev. B **98**, 195410 (2018)

[23] T. Mueller and E. Malic, npj 2D Materials and Applications **2**, 29 (2018)

[24] C. Gong, L. Li, Z. Li, H. Ji, A. Stern, Y. Xia, T. Cao, W. Bao, C. Wang, Y. Wang, Z. Q. Qiu, R. J. Cava, S. G. Louie, J. Xia and X. Zhang, Nature **546**, 265 (2017)

Terahertz Frequency Comb in Graphene Field-Effect Transistors

Pedro Cosme[1,2,*]

[1]Instituto de Plasmas e Fusão Nuclear, 1049-001 Lisboa, Portugal
[2]Instituto Superior Técnico, 1049-001 Lisboa, Portugal

Abstract.
Graphene Field-effect transistors (GFETs) are excellent candidates for all-electric, low-power radiation sources and detectors based on integrated circuit technology. In this work, we show that a hydrodynamic instability can be explored (the Dyakonov–Shur instability) to excite the graphene plasmons. The instability can be sustained with the help of a source-to-drain current and controlled with the gate voltage. It is shown that the plasmons radiate a frequency comb in the Terahertz (THz) range. It is argued how this can pave the stage for a new generation of low power THz sources in integrated-circuit technology.

1 Introduction

Terahertz radiation has numerous applications ranging from sensing and imaging to metrology and spectroscopy [1, 2]; in particular terahertz laser (THL) combs play a prominent role within such technology [3, 4].

However, THL generation still faces significant difficulties, particularly in the low-power bench-top cases, therefore, practical solutions towards inexpensive, compact and easy-to-operate lasing devices are desirable. With the advent of graphene plasmonics, new techniques relying in optical pumping have been put forward [5, 6]. Alongside this, the progress in graphene based transistors [7] paved the way to the possibility for all-electrical miniaturised devices for low power radiation emission and detection.

In this work, we exploit a scheme for the generation of coherent THz frequency combs in graphene field-effect transistors, arising from the Dyakonov–Shur (DS) plasmonic instability [8, 9]. The latter can be excited via the injection of an electric current, thus forgoing the necessity of optical pumping. This opens the possibility to the development of an all-electric, low-consumption stimulated THL, capable of operating at room temperature.

2 Plasmonic instability

2.1 Hydrodynamic model

The electronic flow in graphene is suitable to be modelled by in a hydrodynamic description neglecting viscosity, and assuming also that transport is restricted to occur only in on

*e-mail: pedro.cosme.e.silva@tecnico.ulisboa.pt

direction, x [10, 11]. This motion is governed by the Euler equations

$$\frac{\partial n}{\partial t} + \frac{\partial}{\partial x} nv = 0 \quad \text{and} \quad \frac{\partial v}{\partial t} + v\frac{\partial v}{\partial x} = \frac{F}{m^*} - \frac{1}{m^* n}\frac{\partial P}{\partial x} \tag{1}$$

for the density n and collective velocity v with force F, P the pressure and m^* the carriers mass. The applicability of such hydrodynamic treatment of conduction in graphene is justified by the large mean free path, $l > 0.2\,\mu\text{m}$ at room temperature, that assures the ballistic transport approximation. Moreover, throughout this work electrons are considered to be at degenerate Fermi liquid regime insofar that the Fermi level remains bellow the Van Hove singularities, such conditions in graphene at room temperature translate to $0.025\,\text{eV} \ll E_F \ll 3\,\text{eV}$.

Yet, the fact that electrons in graphene behave as massless fermions pose a difficulty to the development of models with explicit dependency on the mass, in our case the nominal Drude mass $m^* = \hbar\sqrt{\pi n_0}/v_F$, where $v_F = 10^6\,\text{ms}^{-1}$ is the Fermi velocity and n_0 the equilibrium carrier density, is used as an effective mass [10, 12]. Then, the pressure term in (1) can be recast as $P = \hbar v_F \sqrt{\pi n^3}/3$, resorting to the usual Fermi liquid pressure. Moreover, considering a monolayer graphene sheet in a field effect transistor (FET) structure, i.e. placed between two metallic contacts, source and drain, and subject to a gate, as schematised in Fig.1a, the electric force in the electrons is dominated by the imposed potential that screens Coulomb interaction between them. In this conditions the applied bias potential U has the contribution of both the parallel plate capacitance $C_g = \varepsilon/d_0$ and the quantum capacitance [13, 14] $C_q = 2e^2\sqrt{\pi n}/\pi\hbar v_F$ as

$$U = en\left(\frac{1}{C_g} + \frac{1}{C_q}\right). \tag{2}$$

However, for carrier densities in the range $n \gtrsim 10^{12}\,\text{cm}^{-2}$, quantum capacity dominates $C_q \gg C_g$ and the potential can be approximated as $U \simeq end_0/\varepsilon$. Thus, the acceleration of the FET electron fluid simply reads

$$\frac{-e\nabla U}{m^*} = -\frac{e^2 d}{m^* \varepsilon}\frac{\partial n}{\partial x}. \tag{3}$$

Therefore, and keeping the pressure term up to first order, the fluid model can be written in a nondimensional form as:

$$\frac{\partial n}{\partial t} + \frac{\partial}{\partial x}(nv) = 0 \quad \text{and} \quad \frac{\partial v}{\partial t} + v\frac{\partial v}{\partial x} + \frac{S^2}{v_0^2}\frac{\partial n}{\partial x} = 0, \tag{4}$$

where v_0 is the electron mean drift velocity along the graphene channel and $S^2 \equiv \frac{e^2 d_0 n_0}{m^* \varepsilon} + \frac{v_F^2}{2}$ can be interpreted as sound velocity of the carriers fluid, as the dispersion relation $\omega = (v_0 \pm S)k$ is akin to a shallow-waters regime; for typical values of graphene properties the ratio S/v_0 scales up to a few tens.

2.2 Dyakonov–Shur instability

The hydrodynamic model (4) describes an instability [8, 15] under the boundary conditions of fixed density at source $n(x = 0) = n_0$ and fixed current density at the drain $n(x = L)v(x = L) = n_0 v_0$. Such Dyakonov–Shur instability rises from the multiple reflection of the plasma waves at the boundaries whilst being amplified by the driven current at the drain. The imposed drain direct current guarantees the necessary Doppler shift to the upstream current interfere positively with the downstream current .

According to (4) and the referred boundary conditions the excitation complex frequency $\omega = \omega_r + i\gamma$ can be derived, retrieving [8, 16]

$$\omega_r = \frac{|S^2 - v_0^2|}{2LS}\pi \quad \text{and} \quad \gamma = \frac{S^2 - v_0^2}{2LS} \log\left|\frac{S + v_0}{S - v_0}\right|. \tag{5}$$

Therefore, given the dependence of S with gate voltage, and as $v_0 n_0 = I_{DS}/We$ the frequency can be tuned by the gate voltage and injected drain current, not being solely restricted to the geometric factors of the FET.

3 Radiation emission

The DS plasmonic instability leads to well defined shock waves in the FET channel. One of the most significant implications of such shock waves is that they compel the electrons to not only act collectively but, more importantly, to bunch themselves as the shock profile forces the electronic fluid to concentrate upstream. In that sense the electrons undergo a process with some likelihood to the one in free electron lasers. Then, it is no wonder that the radiation provided by such means exhibits coherence to some extent, indeed the numerical results point to a first order temporal degree of coherence $g^{(1)}(\tau) \gtrsim 0.6$.

Furthermore, the collective oscillation of the bunched electrons is responsible for the THz synchrotron radiation emitted, that can be obtained from the density and current evolution of the fluid.

4 Pulsed stimulation & Frequency Comb

If the excitation of the plasmons is solely achieved by a injection of continuous current at the drain, the instability saturates after a transient time, leading to a continuous wave emission. Although continuous wave is relevant for a number of applications, wide bandwidth frequency combs are quite sought after. Thereby, a scheme was idealised to generate such wave packets.

If one considers that a periodic inversion of polarity is imposed across the channel, in such a way that the time on direct polarisation is enough for the saturation to occur, and then

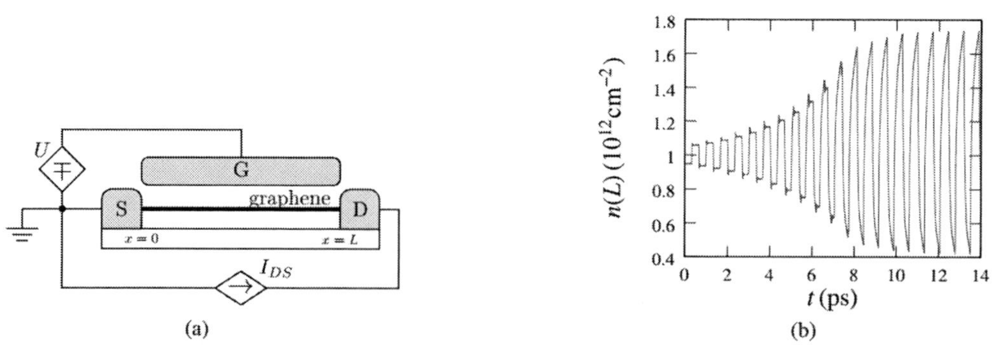

Figure 1. (a) Schematic diagram for gated graphene transistor. For the DS instability to occur a fixed current I_{DS} is injected at the drain while maintaining the electronic density of the source constant. (b) Example of the instability growth of electronic density at the drain.

suddenly inverted, it is easy to see that the plasma instability would be generated – saturate – and then compelled to decay in time. In this way a wave packet is created with an envelope that can be approximated by:

$$A(t) = \begin{cases} e^{\gamma t} - 1 & \text{if} \quad 0 \le t < t_s \\ e^{\gamma t_s} - 1 & \text{if} \quad t_s \le t < t_o \\ e^{-\gamma(t-t_p)} - 1 & \text{if} \quad t_o \le t < t_p \end{cases} \tag{6}$$

where t_s is the characteristic time for saturation, t_o the duration of the excitation and $t_p = t_o + t_s$ the total duration of the pulse, and that can be seen in Fig.2a. The amplitude envelope Fourier transforms into

$$\hat{A}(\omega) = \gamma \sqrt{\frac{2}{\pi}} \frac{e^{t_s \gamma} \omega \cos\left[\frac{\omega}{2}(t_o - t_s)\right] - \omega \cos\left[\frac{\omega}{2} t_p\right] + e^{t_s \gamma} \gamma \sin\left[\frac{\omega}{2}(t_o - t_s)\right] - \gamma \sin\left[\frac{\omega}{2} t_p\right]}{\omega\left(\gamma^2 + \omega^2\right)}, \tag{7}$$

which defines the bandwidth, and general shape, of the frequency comb. The computed full-width at half maximum (FWHM) as function of the pulse duration, t_o, is shown in Fig2b, from where is also evident that the saturation time t_s does not play a significant role in the value of the FWHM.

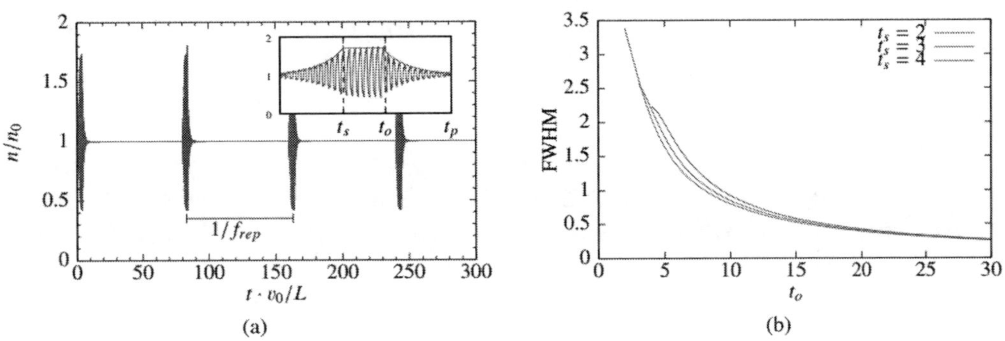

Figure 2. (a) Example of pulsed excitation of DS instability, using a period $100t^*$ with 10% duty cycle. Inset showing the amplitude envelope with the characteristics times. (b) Dependence of FWHM of the spectral envelope from equation (7) with duration of excitation for different saturation times.

The performed simulations showed that the electronic fluid responds as expected to the polarisation inversion originating the pulses that form the frequency comb. In Fig. 3 a paradigmatic example of such frequency combs can be seen, with the central frequency very close to the theoretical value of (5) and the shape of the spectral envelope in accordance with (7), in particular denoting its the non-monotonic silhouette.

5 Conclusion

We make use of a hydrodynamic model to describe a plasmonic instability taking place in a graphene field-effect transistor at room temperature. Our scheme, based on the control of the electron current at the transistor drain, results in the emission of a THz frequency comb. Numerical simulations suggest that the emitted THz radiation is extremely coherent, thus be an appealing candidate for a THz laser source. Our findings point towards a method for the development of tuneable, all-electrical THz antennae, dismissing the usage of external

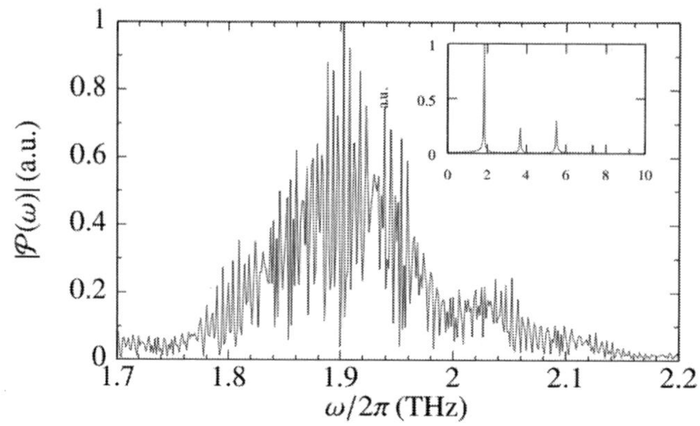

Figure 3. Frequency comb – Total radiated power FFT. Numerical simulation with $L = 0.75\,\mu$m and $v_0 = 0.3v_F$, for constant $S/v_0 = 20$, repetition rate of pulses of 5 GHz with a 10% duty cycle. Inset – Spectrum of the continuous wave case for the same parameters.

light sources . This puts graphene plasmonics and, in particular, graphene field-effect transistors in the run for competitive, low-consumption THz devices based on integrated-circuit technology, allowing the fabrication of patch arrays designed to enhance the total radiated power.

References

[1] D.L. Woolard, W.R. Loerop, M.S. Shur, eds., *Terahertz sensing technology Volume 2: Emerging Scientific Applications & Novel Device Concepts* (World Scientific, 2003), ISBN 9812386114

[2] N. Pornsuwancharoen, M. Tasakorn, P.P. Yupapin, S. Chaiyasoonthorn, Optik **124**, 446 (2013)

[3] H. Füser, M. Bieler, Journal of Infrared, Millimeter, and Terahertz Waves **35**, 585 (2014)

[4] I. Barmes, S. Witte, K.S.E. Eikema, Nature Photonics **7**, 38 (2013)

[5] G. Altares Menendez, B. Maes, Physical Review B **95**, 144307 (2017)

[6] T. Otsuji, A. Satou, M. Ryzhii, V. Popov, V. Mitin, V. Ryzhii, Journal of Physics: Conference Series **486**, 012007 (2014)

[7] F. Schwierz, Nature Nanotechnology **5**, 487 (2010)

[8] M. Dyakonov, M. Shur, Physical Review Letters **71**, 2465 (1993)

[9] Z. Kargar, T. Linn, C. Jungemann, Semiconductor Science and Technology **33** (2018)

[10] A.J. Chaves, N.M. Peres, G. Smirnov, N. Asger Mortensen, Physical Review B **96**, 1 (2017)

[11] M. Müller, J. Schmalian, L. Fritz, Physical Review Letters **103**, 025301 (2009)

[12] D. Svintsov, V. Vyurkov, V. Ryzhii, T. Otsuji, Physical Review B **88**, 245444 (2013)

[13] T. Fang, A. Konar, H. Xing, D. Jena, Applied Physics Letters **91**, 092109 (2007)

[14] S. Dröscher, P. Roulleau, F. Molitor, P. Studerus, C. Stampfer, K. Ensslin, T. Ihn, Physica Scripta **152104**, 014009 (2012)

[15] M. Dyakonov, Comptes Rendus Physique **11**, 10 (2011)

[16] A.P. Dmitriev, A.S. Furman, V.Y. Kachorovskii, G.G. Samsonidze, G.G. Samsonidze, Physical Review B **55**, 10319 (1997)

Modelling isolated hydrogen impurity in Lu_2O_3 with muonium spectroscopy

Rui C. Vilão[1,*], *Ricardo* B. L. Vieira[2], *Helena* V. Alberto[1], *João* M. Gil[1], *Alois* Weidinger[3], *Roger* L. Lichti[4], *Patrick* W. Mengyan[4,5], *Britany* B. Baker[6], and *James* S. Lord[7]

[1]CFisUC, Department of Physics, University of Coimbra, R. Larga, P-3004-516 Coimbra, Portugal
[2]CICECO - Aveiro Institute of Materials, Department of Chemistry, University of Aveiro, 3810-193 Aveiro, Portugal
[3]Helmholtz-Zentrum Berlin für Materialien und Energie, 14109 Berlin, Germany
[4]Department of Physics, Texas Tech University, Lubbock TX 79409-1051, USA
[5]Department of Physics, Northern Michigan University, Marquette, MI, 49855
[6]Department of Physics and Engineering, Francis Marion University, Florence, SC 29502, USA
[7]ISIS Facility, Rutherford Appleton Laboratory, Chilton, Didcot, Oxon OX11 0QX, United Kingdom

Abstract. We identified in this experiment two muon configurations in Lu_2O_3, the oxygen-bound ($O{-}Mu^+$) ground state and a metastable (energy barrier 0.7(3) eV) atom-like excited state. These configurations are partially not formed immediately after implantation but somewhat delayed due to the requirement of a lattice rearrangement around the muon. These rearrangement processes occur on a timescale of ns to μs and are thus observable in μSR experiments. A special role plays a fairly long-lived (ns to μs) transition state as an intermediate step in the reaction process.

1 Introduction

Oxides, including rare earth oxides, are currently under intense scrutiny in the context of applications in electronic devices [1–3]. The hydrogen impurity can have a prominent impact in the respective properties, particularly in the electrical transport [4]. Hydrogen concentrations as large as 1% can be reached, even for samples grown with high quality methods [5, 6].

The knowledge of the isolated hydrogen configurations inside the material is an essential step for the understanding of the electrical properties [7, 8]. Isolated hydrogen is however very difficult to tackle experimentally, and most information has been obtained by using muonium as a light pseudo-isotope [9]. The electronic properties of this bound state of a positive muon and an electron mimic very closely those of hydrogen, and the spectroscopy of the muon spin using implanted muons presents the further advantage of allowing to probe the ground state as well as metastable configurations (usually implied in high-temperature dynamical behaviour).

Despite their growing relevance, not much is known about the hydrogen or muonium configurations in rare-earth oxides. Thorough *ab-initio* calculations have been performed for the hydrogen impurity in the diamagnetic lutetium sesquioxide Lu_2O_3 [10], which have prompted a close inspection of the corresponding experimental muonium configurations.

*e-mail: ruivilao@uc.pt

2 Experimental details

A polycrystalline Lu_2O_3 sample used in the present study was obtained commercially from Alfa-Aesar (REActon 99.995%). The μSR measurements were performed partially at TRIUMF (Vancouver, Canada) [11] and partially at ISIS (UK) [12, 13]. At TRIUMF, high-transverse field (up to 7 T) measurements were performed at the HiTime spectrometer on the M15 surface muon channel. At ISIS, low transverse field (10 mT) were obtained with the EMU spectrometer in the temperature range from 8 K to 700 K. Details of the instruments and of the μSR method can be found on the web pages of TRIUMF and ISIS [11, 12]. Data were analyzed with WiMDA [14].

3 Experimental results and discussion

Measurements in high-transverse field (Fig. 1) reveal the presence of two oscillating components of the muon spin polarization, at the muon Larmor frequency $\nu_d = \gamma_\mu B$ (corresponding to muons with a diamagnetic surrounding), and at the frequency ν_{12} corresponding to a muonium configuration with an isotropic hyperfine parameter A = 3629(2) MHz.

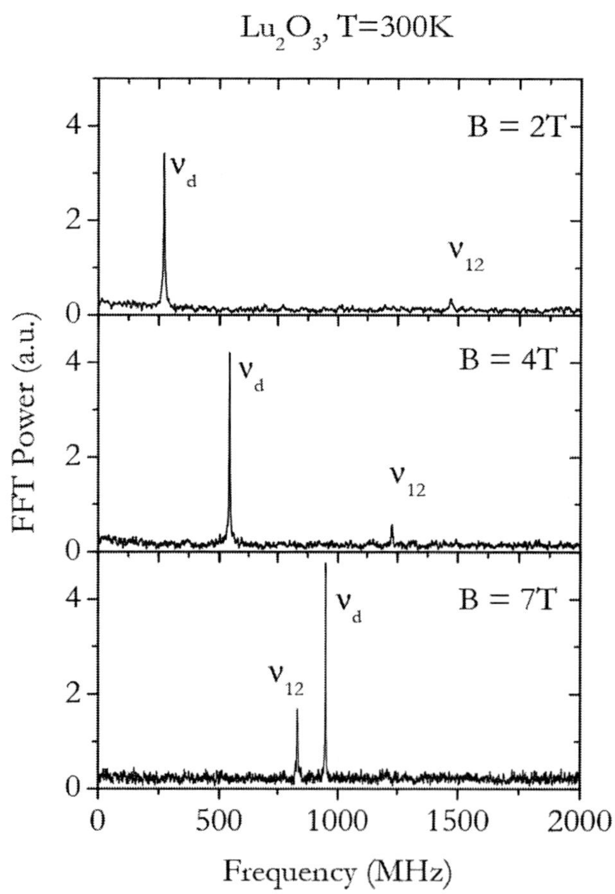

Figure 1. Fast-Fourier transform of high transverse-field μSR measurements at several applied magnetic fields. Two oscillation frequencies ν_d and ν_{12} are clearly seen, corresponding to a diamagnetic state and to a muonium state, respectively.

3.1 Oxygen-bound and atom-like muonium

We will shortly describe the two observed states:

Oxygen-bound muonium. The diamagnetic signal in this experiment is assigned to the configuration where the muon is bound to an oxygen atom. It corresponds to the well-known O–H donor configuration in its ionized state. This configuration was also predicted in theoretical calculations as the ground state of H^+ in Lu_2O_3 [10].

Atomic muonium. The observed state with the large and isotropic hyperfine interaction is assigned to atom-like muonium at an open interstitial site. In theoretical calculations several such sites where identified with similar formation energy and hyperfine interaction [10]. The ground-state configuration is assigned to H^0 at an interstitial unoccupied oxygen site. This state is only metastable and converts to the oxygen-bound configuration at higher temperature (see below).

3.2 Formation probabilities

The temperature dependence of these two components was followed up to room temperature in the high-field measurements; and up to $T = 700$ K in measurements at low transverse field (0.01 T) of the diamagnetic component only (Fig. 2). A transient component, not shown here, has also been identified. It corresponds to a weakly bound (μ^+e^-) configuration which acts as precursor (transition state) for the formation of the final states [15].

Muons are implanted into the target and thus react with the matrix atoms during the slowing down process. They may pick up an electron from the host and form neutral muonium or remain in the positive charge state. During the flight there is not enough time for a rearrangement of the host atoms, thus the muon in either the neutral or the positive charge state stops

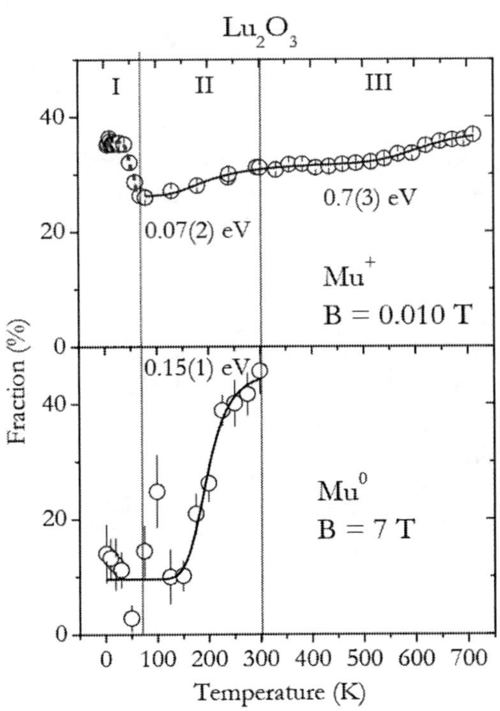

Figure 2. Temperature dependence of the fraction of muons stopping in the (slowly-relaxing) diamagnetic configuration Mu^+ (top) and in the paramagnetic configuration Mu^0 (bottom). The red dashed line is a fit to a "thermal spike model" (see below and Ref. [17]). The black lines are fits to Boltzmann functions as discussed in the text.

in the pristine lattice (Mu^{0*} or Mu^{+*}, the asterisk indicating the excited configuration). The slowing down occurs in the sub-ps time range. The subsequent reactions with the lattice are schematically indicated in Fig. 3.

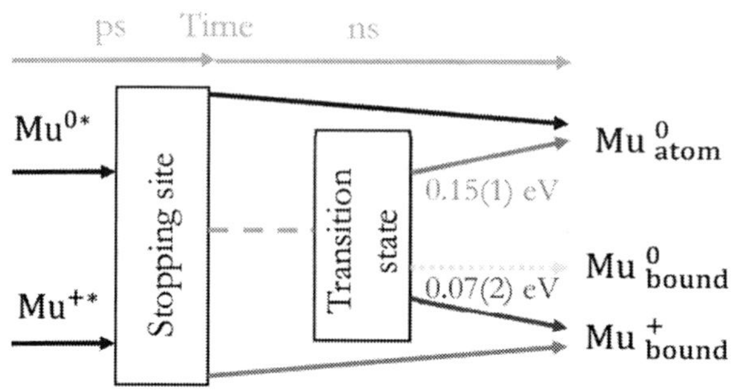

Figure 3. Diagram of the muon formation model based on the presence of a transition state, adapted to the Lu_2O_3 case and including the possible energy barriers for the formation of the final configurations.

The stopped Mu^{0*} and Mu^{+*} may directly go to the corresponding ground states Mu^{0}_{atom} and Mu^{+}_{bound} (black and green arrow in Fig. 3). However, we recently found [16], that there is also the possibility that a fairly long-lived (ns to μs) transition state is formed from which the final configurations arise (blue and red arrow). The transitions are hindered by barriers (energy values on the arrows in the Fig. 3) and are possible only at higher temperatures. This leads to the interpretation of the different temperature regions in Fig. 2.

Region I. The increase of the diamagnetic fraction below about 50 K is a special effect not always seen in μSR experiments. It is due to a local heating of the lattice by the stopping process and by stress release due to relaxation. We call it a thermal spike since it disappears quickly in particular if the temperature is increased. This excitation energy imitates a higher temperatures and can induce reactions which otherwise occur only at higher temperature [17].

Region II. The increase of both fractions with temperature, for the interval between 100 K and 300 K, observed in Fig. 2, has been analysed with phenomenological Boltzmann functions (Ref. [18]), which provide an estimate for the activation energies associated to the corresponding processes: $E_a^+(\mathrm{II}) = 0.07(2)$ eV for the increase of the diamagnetic fraction in region II and $E_a^0(\mathrm{II}) = 0.15(1)$ eV for the increase of the paramagnetic fraction in region II. We assign these reactions to the conversion of the transition state to the oxygen-bound muon (red line in Fig. 3) and the atomic muonium (blue line in Fig. 3), respectively.

Region III. A second step in the diamagnetic Mu^+ fraction is observed in the temperature range between 300 K and 700 K (Fig. 2 top). We assign it to the ionization of ground state muonium, possibly via intermediate states, such as the oxygen-bound configuration with subsequent loss of the electron. The activation energy is $E_a = 0.7(3)$ eV.

4 Conclusions

We observe in this μSR experiment on Lu_2O_3 two different states, one with a diamagnetic signature which we assign to the oxygen-bound (O–Mu^+) ground state configuration, and one with an atom-like characteristic assigned to muonium at an open interstitial site. This later state is metastable and converts at higher temperatures to the oxygen-bound configuration (E_a = 0.7(3) eV). The embedding of the muon into the host lattice requires a rearrangement of the surrounding atoms. This takes some time (ns to μs) and makes intermediate steps observable in μSR. We observed in particular i) an effect from the deposited energy at the end of the muon trajectory (thermal spike) which initiates reactions occurring otherwise only at higher temperatures, and ii) the delayed conversion from the intermediate transition state to the final configurations. Such delayed processes occur also in other implantation processes but are usually not observable. The muon as a sensitive local probe makes them visible.

Acknowledgements

Access to the ISIS and TRIUMF beams, as well as the support of the ISIS and TRIUMF muon teams are gratefully acknowledged. This work was supported with funds from FEDER (Programa Operacional Factores de Competitividade COMPETE) and from FCT - Fundação para a Ciência e Tecnologia (Portugal) under projects UID/FIS/04564/2016 and PTDC/FIS-MAC/29696/2017.

References

[1] E. Fortunato, P. Barquinha, and R. Martins, Advanced Materials **24**, 2945 (2012).

[2] L. Petti, N. Münzenrieder, C. Vogt, H. Faber, L. Büthe, G. Cantarella, F. Bottacchi, T. D. Anthopoulos, and G. Tröster, Applied Physics Reviews **3**, 021303 (2016).

[3] R. Gillen and J. Robertson, Microelectronic Engineering **109**, 72 (2013).

[4] C. G. Van de Walle and J. Neugebauer, Annu. Rev. Mater. Res. **36**, 179 (2006).

[5] R. Alvero, A. Bernal, I. Carrizosa, J. A. Odriozola, and J. M. Trillo. Journal of the Less Common Metals **110**, 425 (1985).

[6] J. Päiväsaari, M. Putkonen, and L. Niinistö, Thin Solid Films **472**, 275 (2005).

[7] A. Weidinger, J. M. Gil, H. V. Alberto, R. C. Vilão J. Piroto Duarte, N. Ayres de Campos, and S. F. J. Cox, Physica B: Condensed Matter **326**, 124 (2003).

[8] R. L. Lichti, K. H. Chow, J. M. Gil, D. L. Stripe, R. C. Vilão, and S. F. J. Cox, Physica B: Condensed Matter **376-377**, 587 (2006).

[9] S. F. J. Cox, R. L. Lichti, J. S. Lord, E. A. Davis, R. C. Vilão, J. M. Gil, T. D. Veal, and Y. G. Celebi, Physica Scripta **88**, 068503 (2013).

[10] E. L. da Silva, A. G. Marinopoulos, R. B. L. Vieira, R. C. Vilão, H. V. Alberto, J. M. Gil, R. L. Lichti, P. W. Mengyan, and B. B. Baker, Phys. Rev. B **94**, 014104 (2016).

[11] More information at the webpage of the Centre for Molecular and Materials Science at TRIUMF: http://musr.ca/.

[12] More information at the webpage of the Muon Group at the ISIS Facility: https://www.isis.stfc.ac.uk/Pages/Muons.aspx.

[13] R. C. Vilão, J. S. Lord, J. M. Gil, H. V. Alberto, and J. Piroto Duarte, STFC ISIS Neutron and Muon Source (2012), https://doi.org/10.5286/ISIS.E.24089275.

[14] F. L. Pratt, Physica B (Amsterdam) **289–290**, 710 (2000).

[15] R. C. Vilão, R. B. L. Vieira, H. V. Alberto, J. M. Gil, A. Weidinger, R. L. Lichti, P. W. Mengyan, B. B. Baker, and J. S. Lord, Phys. Rev. B **98**, 115201 (2018).

[16] R. C. Vilão, R. B. L. Vieira, H. V. Alberto, J. M. Gil, and A. Weidinger, Phys. Rev. B **96**, 195205 (2017).

[17] R. C. Vilão, H. V. Alberto, J. M. Gil, and A. Weidinger, Phys. Rev. B **99**, 195206 (2019).

[18] R. C. Vilão, A. G. Marinopoulos, R. B. L. Vieira, A. Weidinger, H. V. Alberto, J. P. Duarte, J. M. Gil, J. S. Lord, and S. F. J. Cox, Phys. Rev. B **84**, 045201 (2011).

Investigating the anisotropic compression and high-pressure phase symmetry of orthorhombic $RFeO_3$ vs $RMnO_3$

R. Vilarinho[1,*], *A. Almeida*[1], *and J. Agostinho Moreira*[1,*]

[1]IFIMUP, Physics and Astronomy Department, Faculty of Sciences, University of Porto, Rua do Campo Alegre 687, s/n- 4169-007 Porto, Portugal.

Abstract. Rare-earth orthoferrites and orthomanganites present strong couplings between their structural distortions and physical properties, namely magnetoelectricity, generating interest in predicting and realizing new phases under external parameters, such as hydrostatic pressure. In this regard, we discuss the differences and the similarities of the high-pressure behaviour of the structural distortions arising from anisotropic volume compression. This allows a better understanding of the role played by the octahedra tilt and Jahn-Teller effect as pressure accommodation mechanisms on the stabilization of different crystallographic phases after the insulator to metal structural transition occurring at the critical pressures (Pc) between 35 and 50 GPa.

1 Introduction

The physical properties of perovskite materials, with general formula ABO_3, are known to be strongly dependent on their structural distortions, which can be changed through temperature, applied electric/magnetic fields, and epitaxial strain. The understanding of the delicate energy balance involving the different structural instabilities is therefore a key point for the design of multifunctional perovskite-based materials, and can give important inputs for *ab initio* calculations and phenomenological models [1–3]. Hydrostatic pressure allows modifying the interatomic distances and, thus, tuning the interactions to a much larger extent that any other external parameters. We have amply used this advantage in the past for the understanding of model perovskites [4–6].

Rare-earth orthoferrites ($RFeO_3$), and orthomanganites ($RMnO_3$) (with R lanthanide atom) exhibit interesting magnetic phases. In the $RFeO_3$, ferroelasticity and the predicted ferroelectric properties at room conditions in epitaxial films have renewed the interest for this family of compounds [7,8]. This motivates a better understanding of the compressibilities of these structures. All of $RFeO_3$ and the $RMnO_3$, with R = La to Dy, crystallize at ambient conditions in the *Pnma* space group, characterized by two independent tilts of the oxygen octahedra: the in-phase rotation about the $[010]_{pc}$ axis, and the out-of-phase $[101]_{pc}$ axis. [7]. For the same rare-earth, the $RFeO_3$ and $RMnO_3$ exhibit exactly the same the tolerance factor, as Mn^{3+} and Fe^{3+} have the same ionic radius for the 6th coordination in the high spin configuration. However, the MnO_6 octahedron exhibits

Jahn-Teller distortion, which is superimposed to the intrinsic distortion arising from the out-of-phase $[101]_{pc}$ octahedra tilt.

In order to disclose the effect of the structural distortions referred to above on the physical properties of $RFeO_3$, $RMnO_3$, their high-pressure behaviour has been studied to some extend by x-ray diffraction, Mössbauer spectroscopy, and resistivity measurements [5,6,9–14]. A general scenario can be summarized, wherein both types of compounds undergo compression before a structural phase transition occurs at P_C in the 35 to 50 GPa range, depending on the rare-earth ion. This phase transition, usually of the first-order, with a volume collapse of the order of 5%, is accompanied with an insulator-to-metal transition, which has been explained by a high-spin to low-spin state change of the Fe^{3+} and Mn^{3+} cations by Mössbauer spectroscopy [5,6,9–12].

This work focuses on the similarities and differences found in the high-pressure behaviour of $RFeO_3$ and $RMnO_3$ to evidence the role played by the Jahn-Teller distortion, namely on the degree of anisotropy, volume compression, and symmetry of the high-pressure crystallographic phase.

2 Results and discussion

This part of the work will be mainly based on the experimental results reported by R. Vilarinho et al. for the $RFeO_3$ [6] and by D. Mota et al. for the $RMnO_3$ [5], as they embraced the high-pressure behaviour for a large set of rare-earth ions in these systems. These works were based on the synchrotron x-ray diffraction and Raman scattering techniques. From these experimental results, some physical quantities, which were not presented in the original papers, are now calculated and compared.

2.1 High-pressure phase transition

The insulator-to-metallic phase transition in both $RFeO_3$ and $RMnO_3$ series was revealed by the disappearance of the Raman signature above a certain critical pressure. Simultaneously, changes on the x-ray diffraction patterns are found and on the pressure dependence of the lattice parameters obtained from them. The transition is reversible being both the Raman spectrum and the crystallographic structure recovered upon pressure decreasing, with a significant hysteresis ranging between 3 to 6 GPa depending on the rare-earth cation.

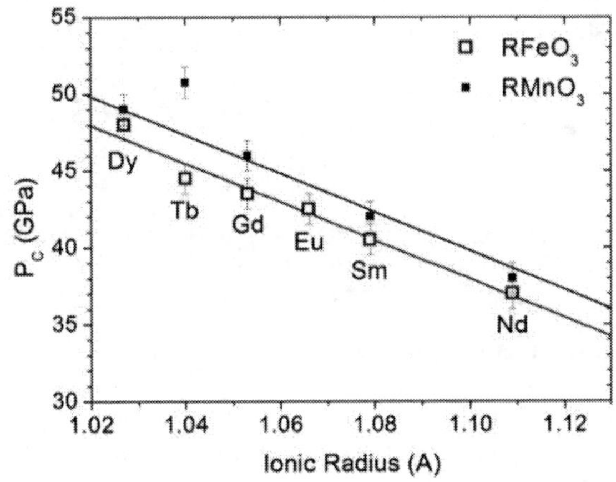

Fig. 1. Critical pressure for $RFeO_3$ (open symbols) and $RMnO_3$ (closed symbols) as a function of the ionic radius for the 6th coordination in the high spin configuration. Data from Refs [5] and [6].

The value for the critical pressure P_C, obtained from the synchrotron x-ray diffraction and Raman scattering data, are depicted in Figure 1 [5,6]. In both cases, as the rare-earth cation size decreases, the critical pressure increases. It is reasonable to assume that, as pressure increases and the FeO_6 octahedra becomes smaller, a critical volume is reached where the electronic repulsion between the oxygen 2p-electrons and the iron e_g-electrons is such, that it is energetically favourable for the latter to pair with the t_{2g}-electrons, avoiding the oxygens 2p-electrons. If we assume this scenario, it is then expectable that for larger R-cations, where FeO_6 octahedra cannot increase their tilts to accommodate the applied pressure, they must reduce their volume and promote the phase transition at a lower critical pressure. Moreover, one can compare these critical pressures with their respective values for the similar pressure-driven phase transition observed in $RMnO_3$ [5]. The critical pressures for $RMnO_3$ are systematically slightly higher than the ones obtained for $RFeO_3$, by an apparent constant value of 2 GPa. This difference can easily be understood using the same logical reasoning as before by taking into account that the Jahn-Teller distortion present in $RMnO_3$ provides an additional pressure accommodation mechanism than in their respective $RFeO_3$ compounds. It is worth to note that the tilt angles for $RFeO_3$ and $RMnO_3$, with the same rare-earth ion, are almost the same, thus allowing for this comparison [15].

Concerning the symmetry of the high-pressure phase, while for $RFeO_3$ it always exhibits the *Pnma* symmetry, no common rule could be ascertained from the experimental results of $RMnO_3$. Though $DyMnO_3$ in fact has an isostructural phase transition at P_C, the best refinements obtained for the high-pressure phase symmetry of $PrMnO_3$ are tetragonal (*I4/mcm*) and of $GdMnO_3$ are cubic (*P2₁3*) [5]. For the case of $SmMnO_3$ it is not conclusive, but it appears not to be *Pnma* [5]. Figure 2 shows the symmetries of these crystallographic phases on a bar-plot, both for $RFeO_3$ and $RMnO_3$.

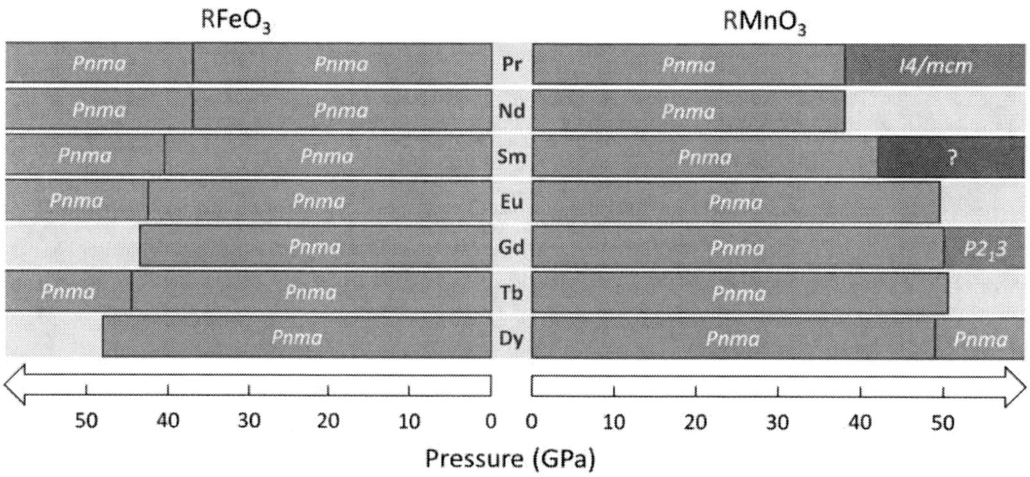

Fig. 2. Symmetry of the crystallographic phases identified in $RFeO_3$ (left) and $RMnO_3$ (right) through x-ray diffraction as a function of applied hydrostatic pressure. Data from Refs. [5,6,9–12].

2.2 Pressure evolution of structural strain

The low quality of the XRD patterns obtained at high pressures has hindered the full Rietveld refinement of the atomic positions and, so, the calculation of the tilt angles. Thus, in order to highlight the role played by the Jahn-Teller distortion and compare on the pressure evolution of the structural distortions below P_C, we have calculated and compared the shear e_4 strain from the lattice parameters for both $RFeO_3$ and $RMnO_3$ [16]:

$$e_4 = [(a_{pc} - a_0) - (c_{pc} - a_0)]/a_0 , \quad (1)$$

where $a_{pc} = a/\sqrt{2}$ and $c_{pc} = c/\sqrt{2}$ are the pseudocubic lattice parameters and $a_0 = \sqrt[3]{V_{pc}}$, with $V_{pc} = a_{pc} x b_{pc} x c_{pc}$. According to Landau theory and symmetry arguments, the symmetry adapted shear e_4 strain provides information regarding primarily the out-of-phase $[101]_{pc}$ octahedra tilt [17].

Figures 3(a) and (b) show the shear e_4 strain as a function of applied pressure for $RFeO_3$ and $RMnO_3$, respectively. For the $RFeO_3$ with rare-earth ions larger than Eu^{3+}, the shear e_4 strain reduces as pressure increases, while it is enlarged for those smaller than Eu^{3+}. $EuFeO_3$ sits exactly at the crossover between these two behaviours, wherein the shear e_4 strain is constant with pressure. This result corroborates the correlation between the shear-strain and the $[101]_{pc}$ octahedra tilt, since the behaviour of the shear-strain with pressure of the $RFeO_3$ exactly mimics that found for the $[101]_{pc}$ octahedra tilt in a previous work [6].

For the $RMnO_3$ a somewhat similar evolution of the pressure dependence of the shear strain on the rare-earth size is found, though the crossover occurs very close to $DyMnO_3$. Thus, for rare-earth ions larger than Dy^{3+}, a reduction of the shear-strain with increasing pressure is observed, while only for $DyMnO_3$ it slightly increases. Two important notes are worth mentioning: i) the rare-earth manganite with the next smaller rare-earth, $HoMnO_3$, is no longer orthorhombic when processed at ambient pressure, and therefore cannot be used for comparison; ii) the $[101]_{pc}$ octahedra tilt cannot be estimated by the lattice parameters alone, as it was done in the $RFeO_3$, due to the higher degree of distortion of the MnO_6 octahedra emerging from the Jahn-Teller effect.

To better elucidate the pressure dependence of the shear strain, Figure 4(a) presents the slopes of the linear pressure-relation of the shear e_4 strain, obtained below 15 GPa, as a function of the rare-earth size for both systems, simultaneously. Although a similar crossover between increasing and decreasing of the shear-strain with increasing pressure is found both in $RFeO_3$ and $RMnO_3$, the evolution of the variation rate with the rare-earth ionic size is much more drastic in the $RMnO_3$, with only $DyMnO_3$ presenting values within those found for $RFeO_3$. From these results we estimate that the crossover between reduction/enhancement of the tilt angles known for $RFeO_3$, not only should also emerge in the $RMnO_3$, but be more pronounced, meaning that the suppression rate of the tilt angles for rare-earth ions larger than Gd^{3+} should be much larger than those found for any of the $RFeO_3$.

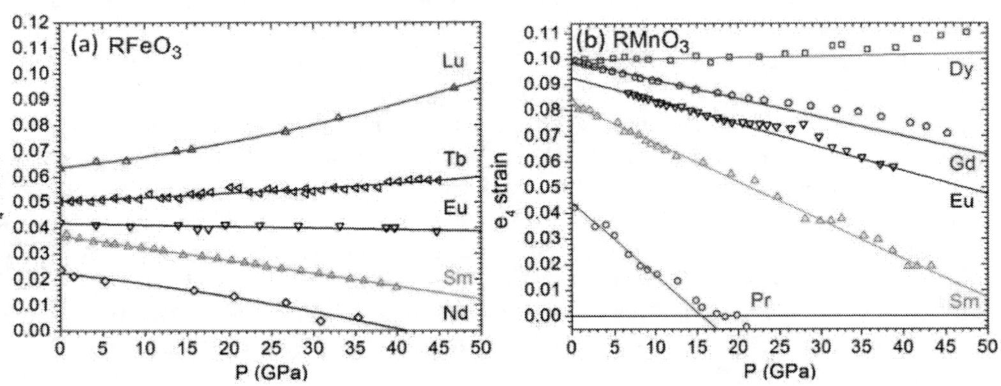

Fig. 3. Shear e_4 strain pressure dependence for (a) $RFeO_3$ and (b) $RMnO_3$. Data of (b) from Ref. [5].

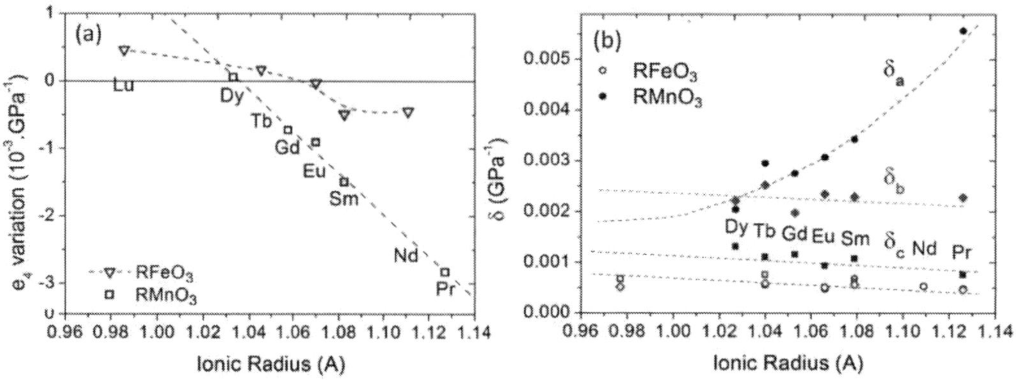

Fig. 4. (a) Shear e_4 strain linear pressure slope and (b) compressibility of lattice parameters for $RFeO_3$ and $RMnO_3$, as a function of rare-earth ionic radius. Data for $RMnO_3$ from Ref. [5].

The apparent change of the shear-strain as pressure increases can be straightforwardly associated to an anisotropic volume reduction. In fact, if the volume reduction was isotropic, the shear-strain would be constant with pressure, as it is the case of $EuFeO_3$. Thus, we have proceeded to calculate the directional compressibilities $\delta_i = 1/(3B^i_0)$, where B^i_0 is the bulk modulus obtained by the best fit of the third-order Birch-Murnaghan equation of state to the lattice parameter *i*. Figure 4(b) presents these compressibilities as a function of the rare-earth ionic radius for both $RFeO_3$ and the $RMnO_3$. In the case of rare-earth orthoferrites (open symbols), the volume reduction is close to isotropic, since the three different compressibilities take the same value. This result agrees with the small variation rates of the shear-strain found for the $RFeO_3$. Conversely, for the rare-earth orthomanganites (closed symbols) this is not the case, where $\delta_a > \delta_b > \delta_c$. The anisotropy of the volume reduction strongly increases as the rare-earth ion size increases, due to the large increase of δ_a. Again, these results agree with the large variation rate of the shear-strain found in the $RMnO_3$, namely for the largest rare-earths.

The different behaviour found on the $RFeO_3$ and the $RMnO_3$ must be attributed to the Jahn-Teller distortion, since it is present in the latter but not in the former. This means that the Jahn-Teller distortion is in fact an important mechanism to accommodate hydrostatic pressure, being also the driving force for the anisotropic volume reduction. Moreover, we can also realize that this distortion is most probably the mechanism yielding the lack of a rule for the high-pressure phase symmetry of the $RMnO_3$, in contrast with the common isostructural one found in the $RFeO_3$. This interpretation is based on the fact that since the volume reduction of the $RFeO_3$ is close to isotropic, even though the octahedra tilts and shear-strain are not constant, the *Pnma* structure is not changed too much, and is still suitably stable to be maintained after the volume collapse at P_C. This explanation even predicts that $DyMnO_3$, the only $RMnO_3$ with variation rates of shear-strain within the range found for $RFeO_3$, would also maintain the *Pnma* symmetry above P_C, which in fact is observed. Contrarily, for the $RMnO_3$ from Gd to larger radius rare-earth ions that present a strong anisotropic volume reduction up to P_C, the *Pnma* structure becomes unstable at P_C, above which the system stabilizes in other more suitable symmetries. Thus, the symmetry of the high-pressure phase is different from the one of the low-pressure phase, and it depends on the degree of anisotropy of the volume reduction up to the transition pressure.

3 Conclusions

In this work, we have compared the structural evolution of $RFeO_3$ and $RMnO_3$ under applied hydrostatic pressure, from which the critical pressure of the insulator-to-metal phase transition was obtained, and the nature of the high-pressure phase symmetry discussed. We were able to conclude that the pressure accommodation in the $RMnO_3$ is much more anisotropic than in the $RFeO_3$, which is almost isotropic. This feature yields the shear-strain to change more drastically in the $RMnO_3$ than in the $RFeO_3$, though for both a crossover between an enhancement and a reduction of this strain occurs for R = Dy and Eu, respectively. The anisotropy of the volume reduction in the $RMnO_3$ may be responsible for the different high-pressure phase symmetries for different rare-earth ions, in contrast with the common isostructural phase transition of the $RFeO_3$. Since the Jahn-Teller distortion marks alone the structural difference between $RFeO_3$ and $RMnO_3$, the degree of the anisotropic compression can be apparently ascribed to the role played by that distortion.

References

1 J. Zhao, N.L. Ross, R.J. Angel, Acta Crystallogr. Sect. B Struct. Sci. **60**, 263 (2004)

2 H.J. Xiang, M. Guennou, J. Íñiguez, J. Kreisel, L. Bellaiche, Phys. Rev. B **96**, 054102 (2017)

3 P. Chen, M.N. Grisolia, H.J. Zhao, O.E. González-Vázquez, L. Bellaiche, M. Bibes, B.G. Liu, J. Íñiguez, Phys. Rev. B **97**, 024113 (2018)

4 J. Oliveira, J. Agostinho Moreira, A. Almeida, V. Rodrigues, M.M. Costa, P. Tavares, P. Bouvier, M. Guennou, J. Kreisel, Phys. Rev. B **85**, 052101 (2012)

5 D.A. Mota, A. Almeida, V.H. Rodrigues, M.M.R. Costa, P. Tavares, P. Bouvier, M. Guennou, J. Kreisel, J. Agostinho Moreira, Phys. Rev. B **90**, 054104 (2014)

6 R. Vilarinho, P. Bouvier, M. Guennou, I. Peral, M.C. Weber, P. Tavares, M. Mihalik, M. Mihalik, G. Garbarino, M. Mezouar, J. Kreisel, A. Almeida, J. Agostinho Moreira, Phys. Rev. B **99**, 064109 (2019)

7 H.J. Zhao, Y. Yang, W. Ren, A. Mao, J. Phys. Condens. Matter 472201 (2014).

8 J.-H. Lee, Y.K. Jeong, J.H. Park, M.-A. Oak, H.M. Jang, J.Y. Son, J.F. Scott, Phys. Rev. Lett. **107**, 117201 (2011)

9 A.G. Gavriliuk, G.N. Stepanov, I.S. Lyubutin, A.S. Stepin, I.A. Trojan, V.A. Sidorov, Hyperfine Interact. **126**, 305 (2000)

10 A.G. Gavriliuk, I.A. Troyan, R. Boehler, M.I. Eremets, I.S. Lyubutin, N.R. Serebryanaya, J. Exp. Theor. Phys. Lett. **77**, 619 (2003)

11 M. Baldini, V. V. Struzhkin, A.F. Goncharov, P. Postorino, W.L. Mao, Phys. Rev. Lett. **106**, 066402 (2011)

12 M.P. Pasternak, W.M. Xu, G.K. Rozenberg, R.D. Taylor, Mater. Res. Soc. Symp. Proc. **718**, D2.7.1 (2002)

13 M. Etter, M. Müller, M. Hanfland, R.E. Dinnebier, Acta Crystallogr. Sect. B Struct. Sci. Cryst. Eng. Mater. **70**, 452 (2014)

14 N.A. Halasa, G. Depasquali, H.G. Drickamer, Phys. Rev. B **10**, 154 (1974)

15 R. Vilarinho, D.J. Passos, E.C. Queirós, P.B. Tavares, A. Almeida, M.C. Weber, M. Guennou, J. Kreisel, J. Agostinho Moreira, Phys. Rev. B **97**, 144110 (2018)

16 M.A. Carpenter, B.A. I., S. Friedrich, Am. Mineral. **86**, 348 (2001)

17 M.A. Carpenter and C.J. Howard, Acta Crystallogr. Sect. B **65**, 134 (2009)

Electrical response of $La_{2/3-x}Li_{3x}TiO_3$ ceramics obtained by spark plasma sintering

J. Silva-Pereira [1], *F.* Guerrero [1], *Y.* Romaguera-Barcelay [1], *L.* Aguilera[2], *R. S.* Silva[3], *J.* Anglada- Rivera[4], *N. C.* Pillajo[5], *A.* Almeida[6], *J.* Agostinho Moreira[6] *and Y.* Leyet [1, 5]

[1]Departamento de Física, Universidade Federal do Amazonas, 69077-000, Manaus, AM, Brasil
[2] Departamento de Química, Universidade Federal do Amazonas, 69077-000, Manaus, AM, Brasil
[3]Departamento de Física, Universidade Federal de Sergipe, 49100-000, São Cristóvão, SE, Brasil
[4]Instituto Federal de Educação Ciência e Tecnologia do Amazonas, 69020-120, Manaus, AM, Brasil
[5] LPMAT, Departamento de Engenharia de Materiais, Universidade Federal do Amazonas, 69077-000, Manaus, AM, Brasil
[6]IFIMUP-IN, Institut of Nanoscience and Nanotechnology, Departamento de Física e Astronomia. Faculdade de Ciências da Universidade do Porto, 4169-007 Porto, Portugal

Abstract: The ceramic system $La_{2/3-x}Li_{3x}TiO_3$ presents as an interesting candidate to be used as an electrolyte in solid-state Li-ion batteries. In this paper the electrical response of the ceramic, $La_{2/3-x}Li_{3x}TiO_3$ with x = 0.11 is reported. $La_{2/3-x}Li_{3x}TiO_3$ nanoparticles were synthesized by high energy milling and sintered by Spark Plasma Sintering from an amorphous phase. After sintering, the samples were structurally characterized by XRD and Raman techniques. Measurements of complex impedance varying frequency from 1 Hz to 10 MHz and temperature from 25 °C to 270 °C were performed. The study of DC conductivity allowed us to find the contributions to the total conductivity, grain, and grain boundary of the samples. From the activation energy values, it was possible to determine the conductive mechanism corresponding to the mobility of Li^+ ions.

1 Introduction

Lithium-based batteries are energy sources that need to be studied and developed, taking into consideration, in addition to their efficiency, public and environmental safety issues. In this respect, organic liquid electrolyte batteries with lithium-ion (Li^+) carriers are flammable and susceptible to auto ignition, which may occur due to internal short circuit caused by physical damage after mechanical impact, heating at ambient temperatures, or internal manufacturing defect [1]. Incidents such as lithium-ion battery fires worry about the safety issue in personal electronics, transport vehicles, and commercial airplanes [2]. For this reason, it has been proposed to replace these types of batteries with electrolytes in the solid-state, since they present greater safety in such devices [3-5].

Some inorganic solid-state electrolytes have high ionic conductivity at room temperature (10^{-3} to 10^{-4} S/cm), as well as excellent chemical stability to the electrodes [1]. Other advantages of these electrolytes are: the reduction of heating of the batteries, longer durability, corrosion elimination, etc. Among the lithium-based solid-state electrolytes, lanthanum lithium titanate with perovskite structure (LLTO) has been extensively studied [6]. It has relatively high intrinsic ionic conductivity "dc" at room temperature, reaching values of up to 10^{-3} S/cm. The high conductivity values obtained from LLTO can be related to its open, stable and rigid crystalline structure, with holes and interstices accessible to the Li^+ ion; also, the small mass and size of the lithium-ion facilitate the electric conduction process [7].

It is a recognized fact that synthesis method is decisive to obtain a material with good properties. Among the most recently used techniques, for the synthesis of materials with a particle size of the order of nanometers, the mechanical activation or high energy grinding (MAE) [8] stands out. MAE has been pointed out as an inexpensive way to produce materials out of balance in the most varied forms. The final milling products are generally nanoscale and are characterized by a large surface area, high defect density, and higher diffusion rates [9].

The Spark plasma sintering (SPS) process offers several advantages over conventional methods, such as ease of operation, high sintering speed, high reproducibility, safety, reliability, as well as high-density values in the samples [10]. Recently the authors of this work reported the use of this method for LLTO samples with x = 0.08 [9].

This work has as main objective to characterize the electrical response of the ceramic system $La_{2/3-x}Li_{3x}TiO_3$ with x = 0.11 obtained by the combination of high energy ball milling and sintering via SPS. On the other hand, will be established a comparison with results previously reported by the authors.

2 Materials and Methods

$La_{2/3-x}Li_{3x}TiO_3$ nanopowders were synthesized by the high energy milling technique. The precursor materials, La_2O_3 (Aldrich - 99.9%), Li_2CO_3 (Aldrich - 99.9%) and TiO_2 (Aldrich - 99.7%) were subjected to a high energy milling process in a planetary mill. The sample was obtained by SPS following the same methodology used in the references [9,11].

The structural characterization of the samples was performed by X-ray diffraction (XRD) using a Rigaku D-max diffractometer in continuous mode from 10 to 90 ^{0} in 2θ and Cukα radiation. Another technique used for this purpose was Raman Spectroscopy using a Jobin-Yvon T64000 spectrometer. Measurements were performed with an Ar + 15 mW laser at room temperature using a 514.5 nm excitation line. Meanwhile, for the electrical characterization of the samples, the Complex Impedance Spectroscopy technique was used. Measurements were performed using a Solartron 1260 Impedance Analyzer. A potential of 0.5 V and a range of temperature and frequency of 25 to 270 ^{0}C and 100 Hz to 10 MHz, respectively, were used as measurement parameters. The structural and electrical characterization was similar to that used in reference [9].

3 Results and Discussion

Figure 1a shows the X-ray diffraction patterns of the ceramics $La_{2/3-x}Li_{3x}TiO_3$ doped with x = 0.11 after sintering and heat treatment. In this figure, a tetragonal structure with a spatial group P4 / mmm was identified as the main phase, similar to that reported by other authors

[12]. Additional peaks corresponding to a secondary phase were identified as $Li_4Ti_7O_{16}$. This structure was reported by BERTAUT in 1953 and there is no cif file of it, which makes refinement difficult. A slight variation in the position and intensity of the diffraction peaks can be reported in comparison with the reference [9] where the sample was studied with x = 0.08 Li^+ ions. This behavior may be motivated by an increase in Li^+ ion concentration in the crystalline structure of the material, which creates a larger number of point defects within the crystal structure of the LLTO.

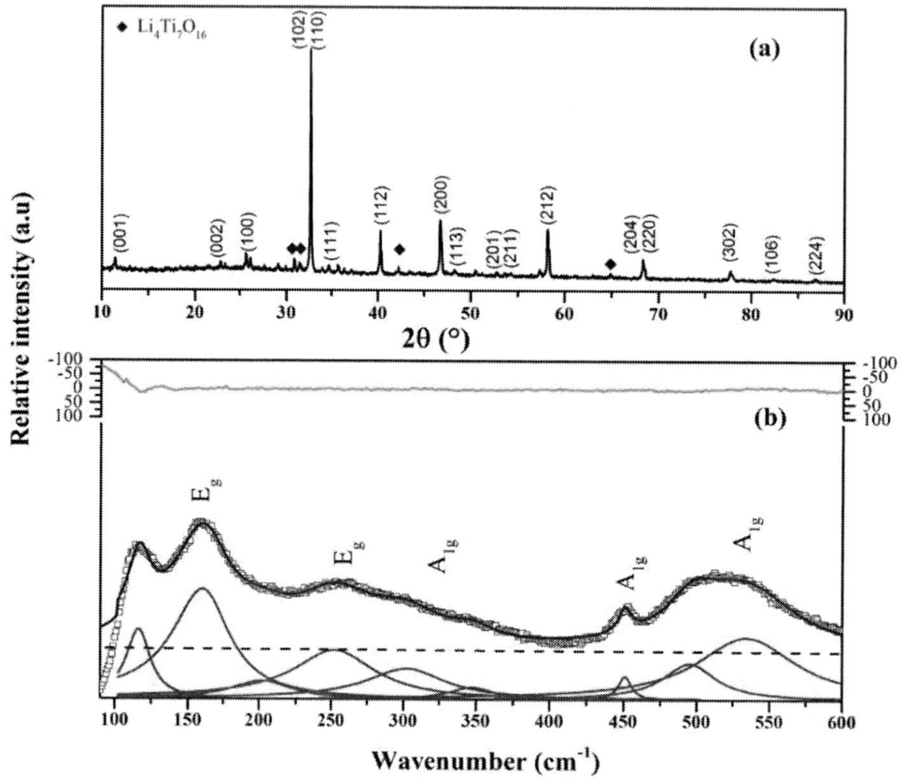

Figure 1. (a) X-ray diffraction patterns of the $La_{2/3-x}Li_{3x}TiO_3$ ceramics obtained by SPS, x = 0.11. The marked peaks (•) correspond to a secondary phase of $Li_4Ti_7O_{16}$. (b) Raman spectra of the $La_{2/3-x}Li_{3x}TiO_3$ ceramic with x = 0.11.

The Li^+ ions insertion into the crystalline structure of the samples was also corroborated by Raman vibrational spectroscopy. The Raman spectra at room temperature of the ceramics system are shown in Figure 1b.

The main Raman bands identified for the sample are similar to those reported in the literature and by the authors of this work [9,13]. The Raman spectra were fitted by oscillator functions with an IGOR PRO 6.0 program, where for 0.11 the bands were located at 163, 256 and 537 cm^{-1} with symmetry Eg were found, while bands 315, 448, 557 and 580 cm^{-1} were detected in the A1g modes. Thus, the band at 163 cm^{-1} is associated mainly with the cations of Ti, more precisely to the vibration of Ti in the plane. The 256 cm^{-1} band is associated with oxygen on the plane. The bands at 305 cm^{-1} are composed mainly of Ti-vibration along the c-axis and the 451 cm^{-1} bands pertained to along the axis c on oxygen (1, 2). The band at 537 cm^{-1} seems more likely to be due to O(3) on the plane. A summary of the observed bands is shown in Table 1. They are obtained from the best fit of the spectra.

The obtained Raman bands are compatible with the tetragonal structure P4/mmm, evidenced by the analysis of the X-ray pattern [9,12,13]. However, it is worth mentioning that Raman spectra present a displacement when compared with previously reported samples [9]. This behavior can be caused by a higher concentration of Li^+ ions in the crystalline structure of the samples; which should contribute to an increase in the number and density of point defects. A similar result was obtained in X-ray diffraction.

Table 1. Values obtained from the Raman spectra adjustments of the LLTO samples with x = 0.08 and x = 0.11.

*x = 0.08**	*x = 0.11*	*Symmetry*	*Atomic displacement*
118	117		
150	163	E_g	Ti - on the plan
178	215		
247	256	E_g	O(3) - on the plan
324	305	A_{1g}	Ti – along the axis c
389	348		
459	451	A_{1g} (prohibited)	O(1,2) – along the axis c
	498		
532	537	E_g	O(3) - on the plane
562		A_{1g} (permitted)	O(3) – along the axis c
589		A_{1g} (permitted)	O(3) – along the axis c

*Results of reference [9]

As it is known, the dopant elements introduction into the crystalline structure of the materials generates some changes in the physical properties. In the case of lithium content, this fact directly affects the structural properties and electrical characteristics of the LLTO system [13,14]. This work corroborates the above effect, showing the influence of the lithium content on the electrical properties of the LLTO system with x = 0.11.

Figure 2 shows the Nyquist plot of the LLTO system with x = 0.08 (white squares) and the sample x = 0.11 (black squares) at room temperature. Comparing the Nyquist plot of these samples similar behavior can be observed, without significant difference in their maximum values of the imaginary part of the impedance, independently of the lithium content. The window of the real part of the impedance also does not present change with the variation of the lithium content. For both cases, the obtained values are of the same order of magnitude. The same behavior for other temperatures was observed, showing that at least under the conditions that these tests were developed; there is no change in the electrical properties and specifically in the conductivity of the sample. Curves for temperature values are not other shown.

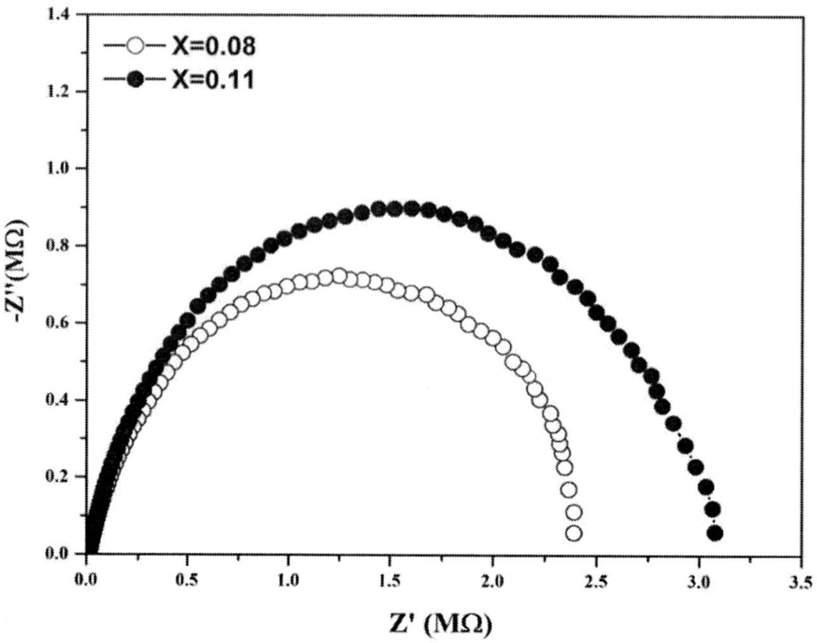

Figure 2: Comparison of the Nyquist plot at room temperature of the LLTO system with X = 0.08 and X = 0.11.

In a similar way to what was done in reference [9] the contributions of grain conductivity, grain boundary, and total conductivity were obtained. The EC (Equivalent Circuit) model was applied to the Nyquist plot of the sample with x = 0.11 to calculate the grain and grain boundary strengths. The circuit proposed to represent the behavior of the sample is the same as that used in reference [9].

Table 2. Conductivity of grain (σ_g), grain boundary (σ_{gb}) and total conductivity (σ_T) obtained via the CE model at different temperatures

T(°C)	σ_g *(S/cm)*	σ_{gb} *(S/cm)*	σ_T *(S/cm)*
25	2,54 x 10^{-05}	1,02 x 10^{-07}	3,33 x 10^{-07}
65	6.17 x 10^{-05}	1.35 x 10^{-07}	1.10 x 10^{-07}
90	1.48 x 10^{-04}	2.08 x 10^{-07}	1.87 x 10^{-07}
120	2.85 x 10^{-04}	5.92 x 10^{-07}	5.42 x 10^{-07}
150	4.5 x 10^{-04}	1.65 x 10^{-06}	1.51 x 10^{-06}
180	6.39 x 10^{-04}	4.01 x 10^{-06}	3.72 x 10^{-06}
210	7.96 x 10^{-04}	9.47 x 10^{-06}	8.78 x 10^{-06}
240	9.33 x 10^{-04}	1.98 x 10^{-05}	1.80 x 10^{-05}
270	1.09 x 10^{-03}	3.98 x 10^{-05}	3.69 x 10^{-05}

From the values of the resistance, calculated by the EC model, and the geometric factor of the sample, it was possible to obtain the values of the dc conductivity of the grain (σ_g) and its grain boundary (σ_{gb}). The total conductivity $\sigma_T = d/R_tS$, where d, R_t and S, are thickness (cm), total electrical resistance (Ω) and area of the sample (cm^2) are calculated from the total resistance, $R_t = R_g + R_{gb}$, respectively. Table 2 shows the values of the conductivity of the grain and grain boundary obtained using the EC model at different temperatures. As can be observed from 25 °C to 270 °C an increase in the dc conductivity values of the sample occurs. A similar behavior was reported in the reference [9].

On the other hand, it was not possible to detect a variation in conductivity values for this sample when compared to that reported in the reference [9]. To a large extent, this result can be strongly motivated by the presence of the secondary phase. This secondary phase reported by X-ray diffraction could be influencing the electrical conductivity of DC and even with the conductive mechanism itself.

The Arrhenius behavior of the sample under study (x = 0.11) is reported in Figure 2, where the values of the activation energies (Ea) of the contribution of the grain, its contour and total conductivities are shown. It is observed the same behavior reported in [9], where the total activation energy is very close to the activation energy of the grain boundary. Therefore, it can be concluded that the grain boundary has a determinant contribution to the results of the total DC conductivity of the sample. The activation energy values obtained for the grain, border, and total conductivities are similar to those already reported [9,12,15-17]. From this result it is possible to conclude that regardless of the influence on the total conductivity of the system, the conduction mechanism remains the same: Li^+ ion mobility is still present in these samples. Thus, the grain frontier governs the electric conduction process, with its more insulating character, which decreases with increasing temperature.

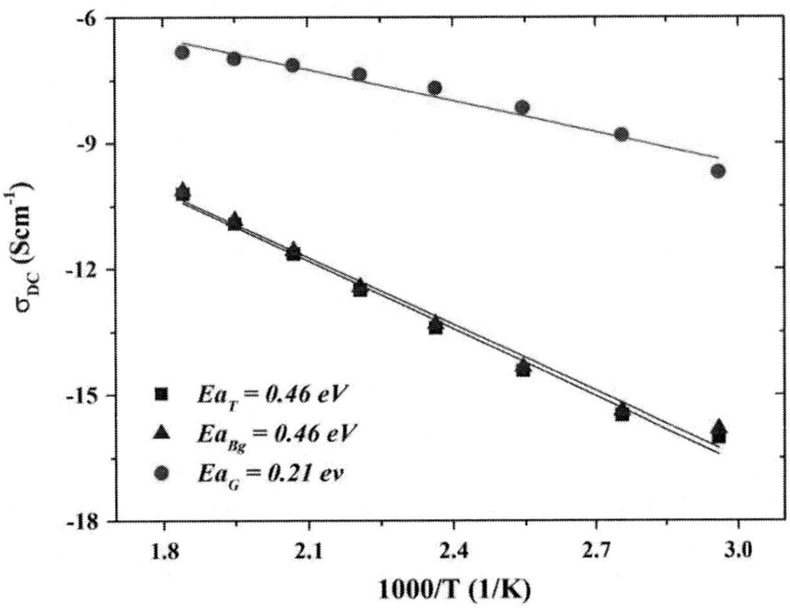

Figure 3. Arrhenius behavior of the grain, grain boundaries and total conductivities of the sample with x = 0.11, sintered via SPS.

4 Conclusions

The LLTO samples with x = 0.11 were made from high energy milling and sintered in the disk form, from a non-crystalline phase, by SPS. The obtained ceramics were characterized structurally, exhibiting a tetragonal phase, with the presence of a secondary phase, corresponding to $Li_4Ti_7O_{16}$. The model of the equivalent circuit was used successfully to obtain the conductivities of each region. From these values, the corresponding DC conductivity values were calculated. In this case, dc conductivity values between 10^{-5} to 10^{-3} S/cm^{-1} and 10^{-7} to 10^{-5} S/cm^{-1} are achieved for grain and grain boundary, respectively. This result evidenced a predominance of the isolating character of grain boundaries of those samples. From Arrhenius plot, the activation energy value for grain (0.214 eV), grain boundaries (0.455 eV), and total conductivity (0.462 eV) was determined. This result corroborates the existence of a conductivity mechanism by diffusion of Li^+ ions, through the crystalline structure of the material. Finally, the sample X = 0.11 did not show considerable improvements compared to the sample x = 0.08 previously reported.

Acknowledgments

The authors would like to thank to CNPq, FAPEAM and CAPES [N° 8454/2014-00] Brazilian and FCT [938/2016] Portuguese agencies for financial support. Finally, Mrs. Inmaculada Martinez, Ph.D Rosalia Poyato, Ph.D Angela Gallardo, for the great help in obtaining of the samples used in this work.

References

[1] F. Zheng, M. Kotobuki, S. Song, M.O. Lai, L. Lu, J. Power Sources **389**, 198 (2018)

[2] B. Zhang, R. Tan, L. Yang, J. Zheng, K. Zhang, S. Mo, Z. Lin, F. Pan, Energy Storage Mater **10**, 139 (2018)

[3] Y. Sun, P. Guan, Y. Liu, H. Xu, S. Li, D. Chu, Crit. Rev. Solid State Mater Sci **44**, 265 (2019)

[4] C. Sun, J. Liu, Y. Gong, D.P. Wilkinson, J. Zhang, Nano Energy **33**, 363 (2017)

[5] S. Chen, D. Xie, G. Liu, J.P. Mwizerwa, Q. Zhang, Y. Zhao, X. Xu, X. Yao, Energy Storage Mater **14**, 58 (2018)

[6] X. Hu, X. Cheng, S. Qin, G. Yan, J. Malzbender, W. Qiang, B. Huang, Ceram. Int. **44**, 1902 (2018)

[7] T. Šalkus, I. Steins, M. Barre, A. Kežionis, A.F. Orliukas, Mater Sci Medziagotyra **19**, 250 (2013)

[8] D. Xie, S. Chen, Z. Zhang, J. Ren, L. Yao, L. Wu, X. Yao, X. Xu, J. Power Sources **389**, 140 (2018)

[9] J.S. Pereira, F. Guerrero, Y. Romaguera-Barcelay, J. Anglada-Rivera, J.C.C. Sales, R.S. Silva, Y. Zulueta, R. Poyato, A. Gallardo, A. Almeida, A. Moreira, Y. Leyet, Mater. Res. Express **6**, 15504 (2019)

[10] H. Zhu, J. Liu, J. Power Sources **391**, 10 (2018)

[11] Y. Leyet, F. Guerrero, J. Anglada-Rivera, I. Martinez, H. Amorin, Y. Romaguera-Barcelay, R. Poyato, A. Gallardo-Lopez, Ferroelectrics **498**, 62 (2016)

[12] T. Yang, Y. Li, C.K. Chan, J. Power Sources **287**, 164 (2015)

[13] M. Romero, R. Faccio, S. Vázquez, S. Davyt, Ceram. Int **42**, 15414 (2016)
[14] Q. Wang, J. Zhang, X. He, G. Cao, J. Hu, J. Pan, G. Shao, J. Eur. Ceram. Soc **39**, 3332 (2019)
[15] H. Geng, A. Mei, Y. Lin, C. Nan, Mater. Sci. Eng. B Solid-State Mater. Adv. Technol **164**, 91 (2009)
[16] T. Šalkus, E. Kazakevičius, A. Kežionis, A.F. Orliukas, J.C. Badot, O. Bohnke, Solid State Ionics **188**, 69 (2011)
[17] C. Uhlmann, P. Braun, J. Illig, A. Weber, E. Ivers-Tiffée, J. Power Sources **307**, 578 (2016)

EPJ Web of Conferences **233**, 05001 (2020) https://doi.org/10.1051/epjconf/202023305001
CMPNC 2019

Ion beam induced current analysis in GaN microwires

Dirkjan Verheij[1,2,1], *Marco* Peres[2], *Susana* Cardoso[1], *Luís Cerqueira* Alves[3], *Eduardo* Alves[2], *Cristophe* Durand[4], *Joël* Eymery[5], *Jorge* Fernandes[6], and *Katharina* Lorenz[1,2]

[1]Instituto de Engenharia de Sistemas e Computadores - Microsistemas e Nanotecnologia (INESC-MN), Rua Alves Redol 9, 1000-029 Lisboa, Portugal
[2]IPFN, Instituto Superior Técnico (IST), Campus Tecnológico e Nuclear, Estrada Nacional 10, 2695-066 Bobadela LRS, Portugal
[3]C2TN, Instituto Superior Técnico (IST), Campus Tecnológico e Nuclear, Estrada Nacional 10, 2695-066 Bobadela LRS, Portugal
[4]CEA INAC-Pheliqs-NPS, Université Grenoble Alpes, Grenoble, France
[5]CEA INAC-MEM-NRS, Université Grenoble Alpes, Grenoble, France
[6]Instituto de Engenharia de Sistemas e Computadores – Investigação e Desenvolvimento (INESC-ID), Rua Alves Redol 9, 1000-029, Lisboa, Portugal

Abstract. GaN is a wide bandgap semiconductor which is expected to withstand high radiation doses. Consequently, it is considered a promising material for new generation particle detectors in radiation related applications. We report on the fabrication and electrical characterization under proton irradiation of single microwire sensors based on a back-to-back Schottky contact configuration. The microwires are grown by metal-organic vapor phase epitaxy and processed into sensors by using optical lithography on dispersed wires. We investigate the impact of the contacts and the semiconductor bulk on the ion beam induced current (IBIC) by irradiating specific areas of the sensor and simultaneously measuring the change in conductivity. We observed that the contribution of the excess charge carriers generated in the depletion regions formed at the contact interfaces is of low influence when compared to the excess charge carriers generated in the microwire bulk.

1 Introduction

The development of radiation resistant electronics and sensors is fundamental to enable technology that can endure extreme harsh environments such as those encountered in space or nuclear facilities. One measure to improve the lifetime of devices is using radiation resistant materials like GaN. GaN is a wide bandgap semiconductor with a good radiation hardness due to its high average displacement energy values, the energy required to displace an atom from its lattice position [1]. Additionally, the radiation defects are mobile even at low temperature, leading to dynamic annealing during the irradiation [2].

1 Corresponding author: dirkjanverheij@ctn.tecnico.ulisboa.pt

Nonetheless, despite being a matured technique, the heteroepitaxial growth of 2D films still presents issues, caused mainly by the lattice mismatch between GaN and the typical growth substrates, which leads to the presence of a high density of dislocations. Exchanging 2D technology with nano- and microwire processing solved this problem to a great extent. Due to the free lateral surface of the wire, the strain caused by the mismatch is relaxed without creating dislocations or the dislocations are bent to the sidewall surface, very close to the bottom of the wire [3].

The detection of ionizing radiation using GaN microwires was reported previously [4]. During proton irradiation a change in conductivity was measured and transient measurements at fixed bias showed a relative current increase of approximately 40%. A strong degradation of the sensor, rendered as an increase of the resistance, was however also observed when the fluence surpassed $1x10^{16}$ protons/cm^2. Also, the presence of a persistent ionocurrent was observed with decay transients of the same order of magnitude as the persistent photocurrent (i.e. current induced by UV light) measured prior to proton irradiation. Persistent ionocurrents were previously observed as well in ZnO nanowires [5]. In GaN wire structures, the persistent photocurrent is attributed to surface recombination. Due to the surface band bending, holes are pushed to surface and in order to recombine, electrons must overcome the corresponding potential barrier [6].

In the present study, single GaN microwire devices are processed by depositing contacts on their extremities. The creation of potential barriers at the interface between the contact metal and the semiconductor leads to a configuration consisting of back-to-back Schottky contacts. We can thus divide the device in three main areas, the forward biased Schottky contact, the reverse biased Schottky contact and the central area of the microwire. The aim of this study is to analyze each of these areas regarding their influence in the electrical current induced by the ion beam irradiation.

2 Experimental methods

GaN microwires were grown on a c-plane sapphire substrate by catalyst-free metal organic vapor phase epitaxy (MOVPE) [7]. To initiate vertical growth, a silane flux needs to be added during the first stage of the growth which leads to a very high incorporation of silicon donor atoms. When the silane flux is turned off halfway the growth process, the Si incorporation decreases. Consequently, the bottom part of the wire is heavily doped (10^{20} cm$^{-3)}$, while the upper part of the wire is moderately doped (10^{18} cm^{-3}) [8,9]. More details on the growth process and characterization of the as-grown microwires can be found in [7]. The length of the wires ranges from 20-25 µm whereas the diameter lies between 1-2 µm. Fig. 1a) shows an SEM image of a single microwire deposited on a Si/SiO_2 substrate.

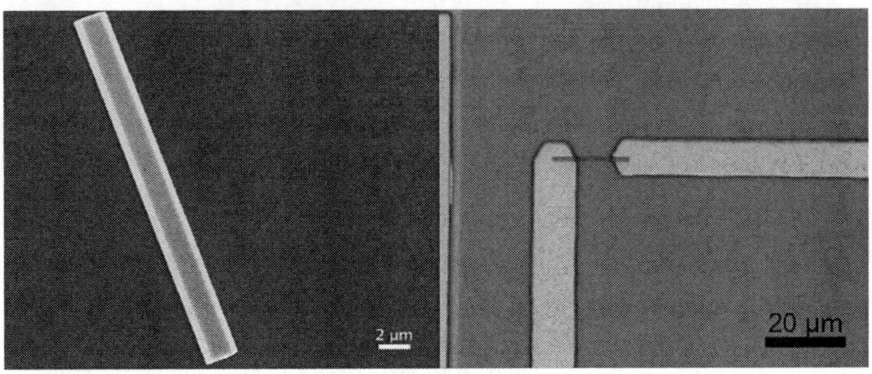

Fig. 1. a) SEM image of a deposited microwire. b) Optical microscopy image of a single wire radiation sensors after processing.

The radiation sensors were fabricated in the form of single microwire devices, with contacts on the extremities of the wires in a similar process as described previously [4]. Prior to the microwire deposition on the device substrate, a metallic pattern containing alignment markers was defined using a standard optical lithography process and a layer of SiO_2 was deposited to avoid conduction between the wires and the substrate. The wires were then dispersed on the SiO_2 using a wet process. A small piece of the wire growth substrate was placed in an Eppendorf tube together with a small volume of isopropyl alcohol (IPA). The tube was placed in an ultrasonic bath for 60 seconds which leads to detachment of the wires from their growth substrate. Finally, a dozen of droplets of the suspension were placed on the device substrate leading, after cleaning the surface with IPA and deionized water, to a uniform distribution of wires. Using the alignment markers, the position of the microwires were defined and metallic Ti/Au (300/4000 Å) contacts were deposited using sputtering. An optical microscope image of a final device is shown in Fig. 1(b).

For the irradiation, hydrogen atoms are ionized and accelerated by a Van de Graaff accelerator to an energy of 2 MeV and directed towards a nuclear microbeam setup [10], where the beam is focused to an area of 3 by 4 μm^2. Taking into account that the beam current is around 70 pA, we have an average ion flux of approximately 3 $x10^{15}$ protons cm^{-2} s^{-1}. The small size of the beam spot combined with the ability to control its position allow the irradiation of specific areas of the device. The electrical properties of the radiation detectors are analyzed prior, during and after the proton irradiation using an Agilent B1500A semiconductor device analyzer. All I-V curves were measured in the dark.

3 Results

Fig. 2 shows the initial current-voltage (I-V) characteristic at room temperature before irradiation. The I-V curve exhibits a non-linear and asymmetric behavior, where the conductivity is higher for positive bias. The non-linearity is a direct consequence of the back-to-back Schottky barriers present at the contacts. The asymmetry is also expected due to the inhomogeneous doping at both extremities of the wire. Since the thickness of the potential barrier is inversely proportional to the square root of the doping concentration [11], the barrier located at the heavily doped extremity has a lower thickness.

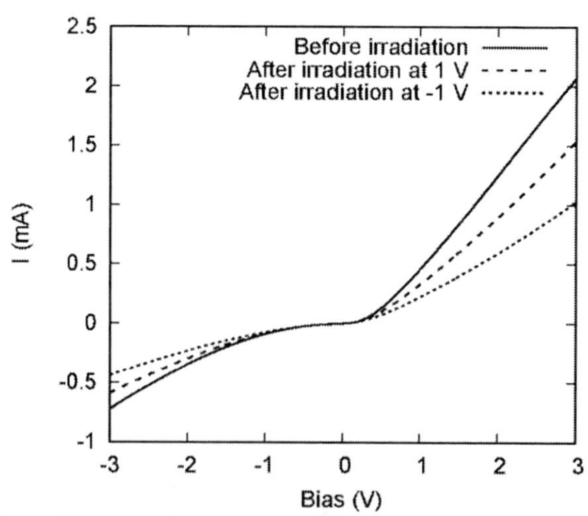

Fig. 2. Experimental I-V characteristics obtained before irradiating the device (solid line) and after irradiating the device (dashed lines). The medium dashed line corresponds to an I-V curve taken after the irradiation at 1 V and the short dashed line ta an I-V curve measured after the irradiation at -1 V.

Therefore, when this contact in reverse bias (corresponding to a positive bias in the graph) the tunneling of electrons through the barrier is enhanced, leading to a higher conductivity and a more linear characteristic. When applying a linear regression to the linear range (1.5 to 3.0 V) we can extract a resistance of $1207.3 \pm 0.5\ \Omega$.

For the irradiation analysis, the sample is placed on a sample holder inside the microbeam chamber, which is evacuated to a pressure of $10^{-5} - 10^{-6}$ mbar. With the aid of proton induced X-ray emission (PIXE) maps the beam is aligned with the sample. Seven areas were defined along the longitudinal direction of the microwire. Area I is located away from the wire, areas II and III are located on the contact with the heavily doped extremity, areas IV and V correspond to the central area of the wire and, finally, areas VI and VII refer to the irradiation at the moderately doped extremity. The beam is placed on each of these areas for a short time instant (~ 2 seconds) in order to minimize the damage and consequent conductivity loss induced by the irradiation. At the same time the current is being measured at a fixed bias. The measurement was done for a bias of +1 Volt, with the contact at the higher doped extremity in reverse bias and repeated for a bias of -1 Volt where the higher doped extremity is forward biased. Due to the back-to-back Schottky configuration of the device, when irradiating the sample, there are two possible mechanisms that can generate ion beam induced current (IBIC). The interaction of the proton beam with the semiconductor leads to the generation of a large number of excess charge carriers due to the band to band excitation. These excess charges are free to move in the semiconductor, therefore, the conductivity of the material increases during the proton irradiation. Beyond this mechanism, near the depletion region at the metal-semiconductor junction, free charge carriers created in the space charge region of the device are separated rapidly by the presence of the internal electric field [12]. In our case, we would expect that the reverse biased Schottky barrier would limit current flow without irradiation and a strong increase of current would be expected upon irradiation. If there was a strong influence of this latter effect, we would observe it when irradiating the areas corresponding to contacts that are reverse biased, as observed by Hirsch *et al.* when performing IBIC measurement on a (Al,Ga)N MSM structure [13]. However, as is shown in fig. 3, the results obtained at positive and negative bias follow the same general behavior. If the contacts would have a significant contribution to the ionocurrent, for positive bias we would see high signals for areas II and III and low signals for areas VI and VII. For negative bias the opposite would happen. Instead, we observe that the IBIC is independent on the polarity of the applied bias.

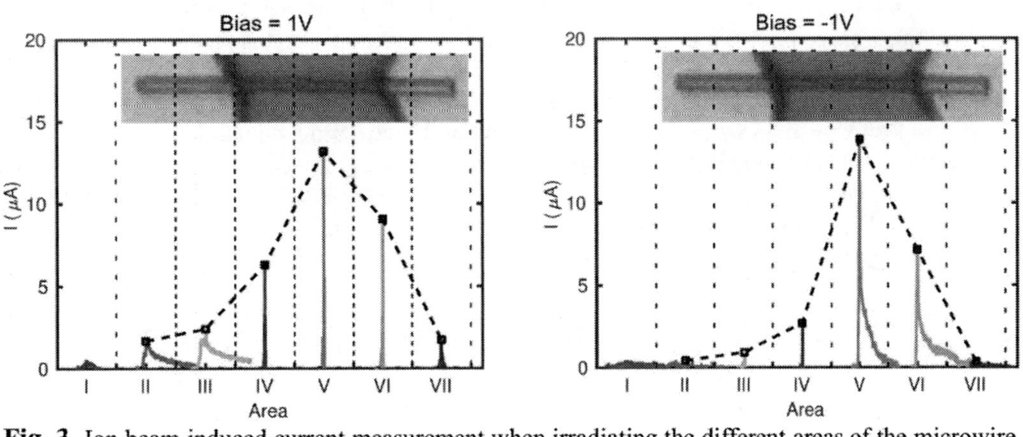

Fig. 3. Ion beam induced current measurement when irradiating the different areas of the microwire, at a bias of 1 V (left) and at a bias of -1 V (right). The dashed lines indicate which area corresponds to each peak. For both positive and negative bias it is clear that the highest IBIC is measured when irradiating area V. Note that for each measurement the background current was subtracted.

This indicates the biggest share of the IBIC is due to free electrons excited to the conduction band in the semiconductor bulk (note that holes in GaN have lower mobility and thus contribute less to the IBIC). A possible explanation for this observation is that, due to the high doping levels, the depletion region, where the free carriers are separated by the internal electric field occurs is too small to provide a significant contribution.

At this moment it is not clear why the IBIC signal is lower in the regions closer to the extremities of the microwires. We note that the alignment of the beam with the microwire is challenging and irradiation may not be completely homogeneous and that the beam intensity can present some instabilities over time. However, assuming that we are irradiating the area corresponding to the extremity of the microwire covered with the metal electrode, the volume of created charge carriers and corresponding IBIC should be of the same magnitude as in the central areas of the wire. Thus, although the experimental results seem to imply that the there is no Schottky effect contribution to the increase of the IBIC, the metal-semiconductor interface seems active in suppressing it. It is possible that this is caused by the different nature of surface defects at the metal-semiconductor interface when compared to the defects present at the semiconductor surface not covered by the metal. The lack of control over the position is however a limitation of the experimental setup making it hard to determine the precise location of the irradiation. Additionally, as mentioned above, the spot size of the beam has an area of 3 by 4 μm^2, which is roughly twice the diameter of the microwire. Consequently, the resolution is relatively small compared to, for example, electron beam induced current (EBIC) measurements [14].

It is also worth looking at the decay of the conductivity due to the damage created by the irradiation. Fig.2 shows the I-V curve measured after each experiment. We can see that the current drops consistently. By performing a linear regression to linear section of the I-V curves we obtain resistances of $1596 \pm 2\ \Omega$ and $2374 \pm 5\ \Omega$, corresponding to the situation after irradiation with a fluence of $(5 \pm 2)\text{x}10^{16}$ protons/cm^2 (after irradiation at positive bias) and a fluence of $(9 \pm 3)\text{x}10^{16}$ protons/cm^2 (after irradiation at negative bias) respectively.

4 Conclusions

In this work the mechanisms of IBIC were studied taking advantage of the characteristics of the microbeam setup allowing irradiation of distinct areas of the device. Although there is some contribution of the Schottky contacts to the I-V curves, the overall influence of the built-in electric field at the Schottky contact on the IBIC signal is small, however we observe a strong suppression of the IBIC in these regions. Although we obtain good stability and reproducible measurements, we observed a decay in conductivity due to the induced irradiation damage. This can have influence on the transient behavior as the creation of traps influences the generation and recombination mechanisms.

In this case however, due to the similar dimensions of the beam and the sample, the resolution of the measurements is not good enough to study the generation processes in depth, nonetheless, it shows there is potential to perform position dependent irradiation measurements for larger size samples.

We acknowledge funding by FCT and FEDER (PTDC/CTM-CTM/28011/2017, LISBOA-01-0145-FEDER-028011)

References

[1] S.J. Pearton, R. Deist, F. Ren, L. Liu, A.Y. Polyakov, J. Kim, Journal of Vacuum Science and Technology A **31**, 050801 (2013)

[2] K. Lorenz, M. Peres, N. Franco, J. G. Marques, S. M. C. Miranda, S. Magalhães, T. Monteiro, W. Wesch, E. Alves, E. Wendler, Proc. SPIE 7940 **7940**, 79400O (2011)
[3] Coulon P M, Mexis M, Teisseire M, Jublot M, Vennéguès P, Leroux M, Zuniga-Perez, J. Appl. Phys. **115**, 153504 (2014)
[4] D. Verheij, M. Peres, S. Cardoso, L. C. Alves, E. Alves, C. Durand, J. Eymery, K. Lorenz, J. Phys. D: Appl. Phys. **51**, 175105 (2018)
[5] A. Johannes, R. Niepelt, M. Gnauck, C. Ronning, Appl. Phys. Lett. **99**, 252105 (2011)
[6] R. Callarco, M. Marso, T. Richter, A. I. Aykanat, R. Meijers, A. v.d. Hart, T. Stoica, H. Lüth, Nano Letters **5**, 981 (2005)
[7] R. Koester, J. S. Hwang, C. Durand, D. L. S. Dang, J. Eymery, Nanotechnology **21**, 015602 (2010)
[8] P. Tchoulfian, F. Donatini, F. Levy, B. Amstatt, P. Ferret, J. Pernot, Appl. Phys. Lett. **102**, 122116 (2013)
[9] P. Tchoulfian, F. Donatini, F. Levy, B. Amstatt, A. Dussaigne, P. Ferret, E. Bustarret, J. Pernot, Appl. Phys. Lett. **103**, 202101 (2013)
[10] N. F. Santos, A. J. S. Fernandes, L. C. Alves, N. A. Sobolev, E. Alves, K. Lorenz, F. M. Costa, T. Monteiro, Nucl. Instrum. Methods Physics Res. B **306**, 195 (2013)
[11] S. M. Sze and K. N. Kwok, Eds., Physics of Semiconductor Devices, 3rd ed. New Jersey: John Whiley & Sons, 2007.
[12] Q. Xu, P. Mulligan, J. Wang, W. Chuirazzi, L. Cao, Nucl. Instrum. Methods Phys. Res. A **849**, 11 (2017)
[13] L. Hirsch, P. Tardy, G. Wantz, N. Huby, P. Moretto, L. Serani, F. Natali, B. Damilano, J.Y. Duboz, J.L. Reverchon, Nucl. Instrum. Methods Phys. Res. B **240**, 265 (2005)
[14] P. Lavenus, A. Messanvi, L. Rigutti, A. De Luna Bugallo, H. Zhang, F. Bayle, F. H. Julien, C. Durand, M. Tchernycheva, Nanotechnology **25**, 255201 (2014)

EPJ Web of Conferences **233**, 05002 (2020) https://doi.org/10.1051/epjconf/202023305002
CMPNC 2019

Micromagnetic study of the vortex state in sub-micron iron discs

Ludgero Peixoto[1,*], *C.* Sousa[1], *D.* Navas[1], and *J. P.* Araújo[1]

[1]IFIMUP and DFA, Faculdade de Ciências da Universidade do Porto, Rua do Campo Alegre 1021/1055, Porto, Portugal

Abstract. Magnetic nanostructures have been widely studied due to its potential applicability into several research fields such as data storage, sensing and biomedical applications. In this work, micromagnetic simulations (mumax3) of sub-micron iron discs are performed for different normalized inter-dot distance (distance/diameter), to better understand the magnetic behaviour of these nanostructures. Two sets of samples were studied: ideal circular discs and disc-shaped nanostructures (based on images of real samples). By analyzing the nucleation and annihilation fields and the magnetic susceptibility, it was found that the (ideal) discs could be considered as isolated for inter-dot distances greater than twice the raidus of the disc ($2R$). The difference in the shape of the disc-shaped nanostructures resulted in an in-plane anisotropy, noticeable on the hysteresis loops for different directions.

1 Introduction

Magnetic nanostructures are, currently, the focus of several research fields, such as data storage, sensing and biomedicine. One subset of these nanostructures are discs that present a magnetic vortex state in remanence. These discs can be used for data storage [1] and for cancer therapy, by magneto-mechanically induced cell death [2]. The vortex state is a direct result of the geometry and aspect ratio of the nanostructure. In the case of the discs, the minimization of energy causes the spins to be in a curling state, gradually changing direction. They start starting parallel to the borders of the disc, remaining in-plane until there is a point where it is energetically favourable for the spins to point either up or downwards, forming the vortex core [3, 4]. This magnetic distribution is more favourable in materials with weak anisotropy [5].

To better understand the influence of the inter-dot distance in the vortex-state, micromagnetic simulations were carried out. These simulations were performed with MuMax3 software [6]. MuMax3 is a GPU-accelerated micromagnetic simulation program that calculates the space- and time-dependent magnetization dynamics in nano-/micro-sized ferromagnets using a finite-difference discretization. The scripting language resembles a subset of the Go programming language. A web-based HTML 5 user interface is provided, allowing the user to inspect and control the simulations from within a web browser [6].

When writting the scripts, firstly, the magnetic constants were defined. The iron magnetic saturation was defined as $M_{sat,Fe} = 1700 \times 10^3$ (A/m), the exchange stiffness $A_{ex,Fe} = 2.1 \times$

*e-mail: ludgero.peixoto@fc.up.pt

10^{-11} (J/m) and the magnetocristalline anisotropy used was $K_{c1,Fe} = 4.8E4$ (J/m^3). The easy axis orientation was also defined as $K_{c1v_{Fe}}$ =vector($cos(45 * pi/180), 0, -sin(45 * pi/180)$) and $K_{c2v_{Fe}}$ =vector($cos(45 * pi/180), 0, sin(45 * pi/180)$). Cell size was chosen to be 5 x 5 x 5 (nm^3), which is smaller than the Fe characteristic length $R_0 = \frac{\sqrt{A}}{M_s}$ =8.5 nm [7], and the damping parameter was taken as 0.5, to ensure rapid convergence. The geometries used were either thin cylinders (circular discs) or the precise shape of the discs taking into account our SEM images, as will be shown. Usually, periodic boundary conditions were set, to better mimic the real sample.

The program starts by issuing a random magnetic state in the desired geometry and taking a snapshot of it, following up by minimizing the energy which results in the fundamental state, also recorded. After this step, the discs are submitted to a magnetic field loop that forces positive and negative saturation, recording the data in a table. At each step, an explicit Runge-Kutta method for advancing the Landau-Lifshitz equation is performed. RK45, the Dormand-Prince method, offers 5-th order convergence and a 4-th order error estimate used for adaptive time step control [6].

2 Ideal discs

For biological applications, not only has the size of the nanostructures to be well-controlled [8], but also is important to understand how they interact with each other. Keeping this in mind, comprehensive simulations were based on circular (ideal) cylinders. The objective was to better understand the interactions between discs with an ideal vortex behaviour, which can be seen in Figures 1 and 2. In these figures two hysteresis loops are shown, corresponding to discs of 300 nm and 50 nm in diameter and thickness, respectively, and with different inter-dot distance. There are two abrupt transitions in the magnetization, which correspond to the nucleation (H_n) and annihilation (H_a) of the vortex in the disc. By monitoring these characteristic fields we could infer about how the inter-dot distance influences the expected vortex behaviour.

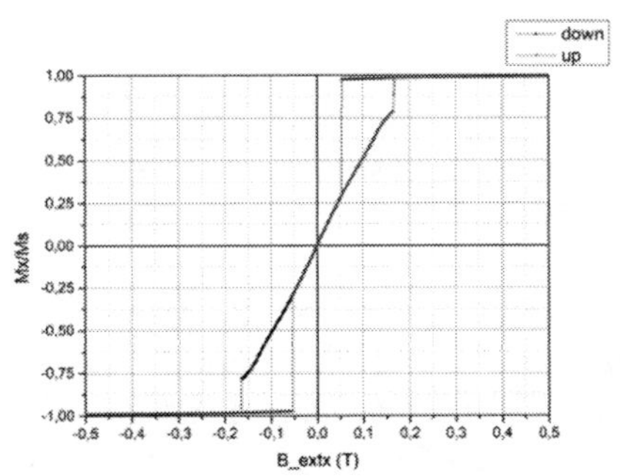

Figure 1: Hysteresis loop for a disc of 300 nm in diameter with 50 nm thickness and inter-dot distance of 35 nm.

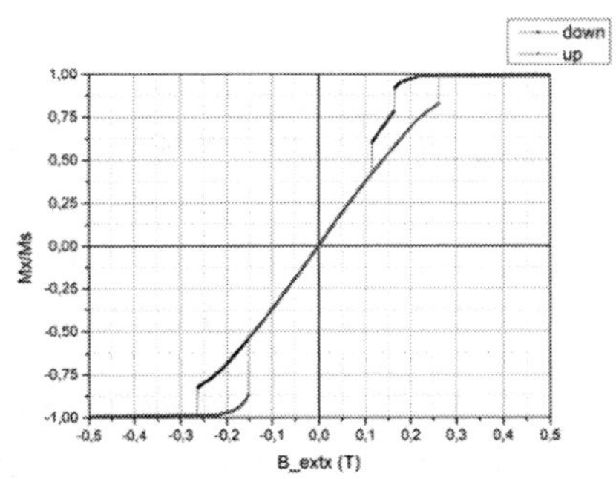

Figure 2: Hysteresis loop for a disc of 300 nm in diameter with 50 nm thickness and inter-dot distance of 300 nm.

2.1 Inter-dot interctions

For studying the inter-dot distance, in more detail, four diameters (200, 300, 400 and 500 nm) were set, for the same thickness (50 nm), varying the inter-dot distance from $R/4$ to just above $2R$, with $24R/10$. The objective here was to assert which disc separation was the threshold to consider a disc isolated. The vortex characteristic fields' dependence with the interdot distance is presented here in Figure 3. The values for the nucleation and annihilation fields, for the same diameter, increase rapidly between $R/4$ to $4R/5$ and then starts to increase more slowly between R and $2R$, before stagnating completely, which is in agreement with a previous work for permalloy nanodiscs [9]. The behaviour can be associated to the increased magnetostatic interactions, for smaller distances, lead to the lowering of the needed field to nucleate a vortex or annihilate it [10, 11]. When the distance becomes large enough, this effect is negligible and the disc can be regarded as isolated. The described behaviour is transversal to all of the four different diameters studied. The threshold seems to be $2R$ but the difference between R and $2R$ is small enough to not be important in situations where a close-packed structure is desirable, as in magnetic memories [1, 12–14].

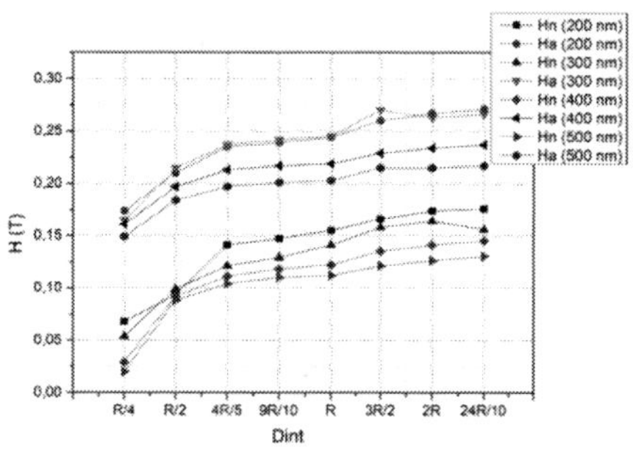

Figure 3: Nucleation (H_n) and anihilation (H_a) fields for different interdot distance (D_{int}).

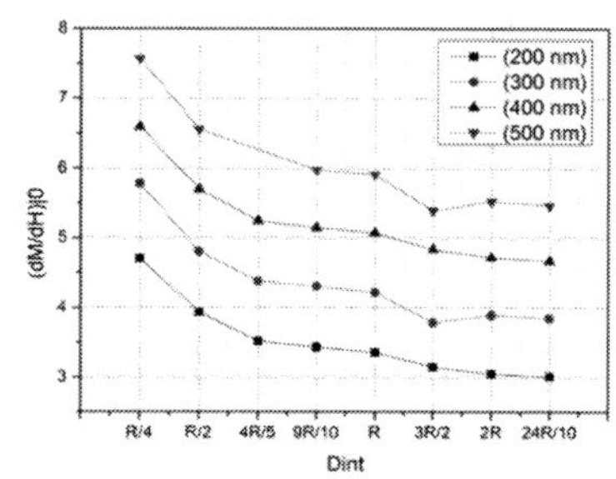

Figure 4: Magnetic susceptibility as a function of interdot distance (D_{int}).

2.2 Magnetic susceptibility

More information can be extracted from analysing the hysteresis loops, as the linear part of the loop can be associated with the vortex core motion [5]. In remanence, when the vortex is (ideally) in the centre of the disc, the moments are aligned with the borders. As the magnetic field is applied, the magnetic moments tend to align with it, disturbing the prior magnetic state and leading to a vortex core motion perpendicular to the applied field. Thus, as the magnetic susceptibility can be related with the derivative of the magnetic moment versus to the field, one can relate it to how easily a vortex core is disturbed and set in motion by an external magnetic field. The dependency on D_{int} represented in Figure 4, can be explained, once again, by the fact that being so close, the dots' magnetic moments cause their neighbours to feel higher fields than the externally applied magnetic field. With more close-packed disc arrays, this effect is accentuated, thus leading to each vortex core needing less applied field to move. As the disc separation increases, the slope value stabilizes, meaning that the disc can be considered not-affected by its neighbours, being an isolated disc [9].

3 Irregular disc-shaped nanostructures

After having dealt with ideal circular discs, the second part of this work revolved around a more practical approach, the simulation of realistically defined disc-shaped nanostructures.

Discs produced by Interference Lithography seem to be good candidates as its viewed as a low-cost technique capable of producing large areas of usable nanostructures. Basing in works using this technique [15], treated SEM images of the discs were used to define the geometry of the scripts. The idea was to replicate as accurately as possible the real sample. The intrinsic anisotropy of the disc meant it would behave differently, even depending on the direction of the in-plane applied magnetic field.

The main objective was, to study and identify typical vortex characteristics in the hysteresis loops when the applied field was parallel to the disc plane (x- and y-direction), and compare it with the ideal case, as illustrated in Figure 5. In both axes, there are two clear, abrupt, transitions corresponding to the vortex nucleation and annihilation, as well as a linear part on the hysteresis loop, representing the movement of the vortex core, as observed in the ideal dics. The vortex characteristic fields were H_n=0,0698 T, H_a=0,129 T for the x-direction and H_n=0,1129 T, H_a=0,1718 T for the y-direction. This is the first and major difference between this case and the ideal one, due to the intrinsic in-plane shape anisotropy, the characteristic fields deppend on the direction of the applied field.

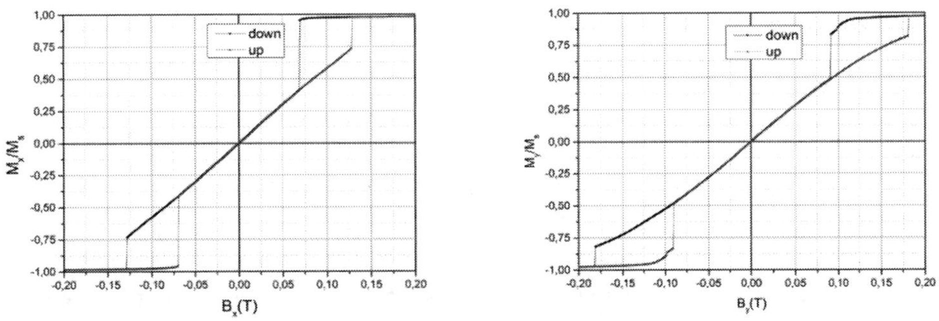

(a) Hysteresis loop in the x-direction.

(b) Hysteresis loop in the y-direction.

Figure 5: Simulated hysteresis loop for single disc with 50 nm of Fe, for in-plane and out of plane applied magnetic field.

Figure 6 shows snapshots at six different applied fields in an array of nine discs, without periodic boundary conditions, to demonstrate the nucleation process. Starting off, all of the discs are in the saturated state (Figure 6a). As the field starts to reduce, some inhomogeneities begin to make an appearance at one of the discs in the outer rows (Figure 6b). This can be explained since no periodic boundary conditions are implemented, therefore the outer discs are the ones are less affected by their neighbours [10, 11]. Between Figure 6c and Figure 6d, all of the disc have already started to nucleate. It is noted that this transition is not as abrupt as presumed, revealing the more probabilistic nature of the event [16]. Even in Figure 6e, some inhomogeneities can be found, before the final image (Figure 6e) where the vortex is present in all of the discs and nearly centred. This images can correspond to the hysteresis loop in Figure 5a.

4 Concluding remarks

The underlying objectives of this study were to better understand the neighbouring discs' interactions and how it affects the vortex characteristic parameters (nucleation, annihilation and

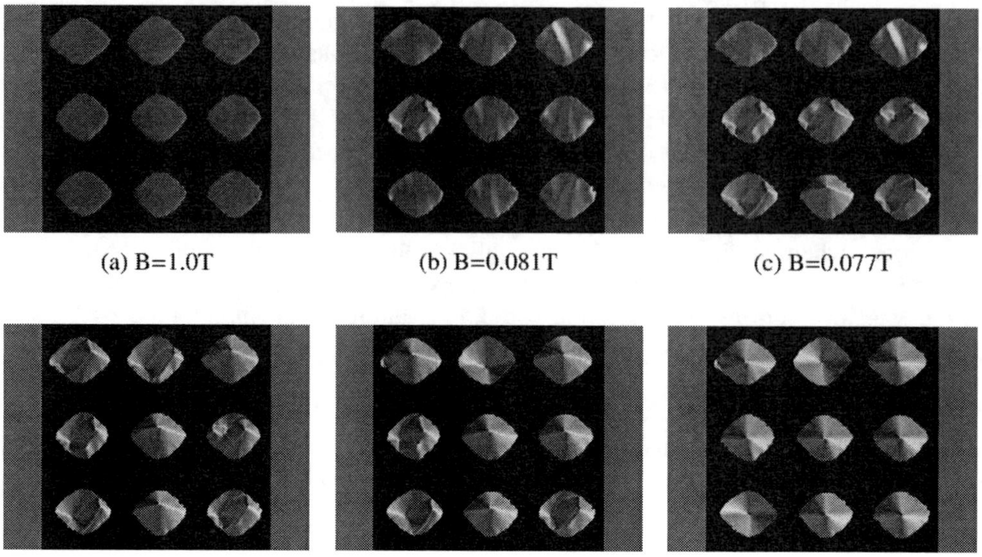

(a) B=1.0T (b) B=0.081T (c) B=0.077T

(d) B=0.067T (e) B=0.056T (f) B=0.011T

Figure 6: Field evolution of an array of 50 nm thickness Fe discs, without PBC

magnetic susceptibility). When the separation between discs was higher than 2R, the interactions were found to be negligible, and so, the disc could be considered isolated. This was consistent between all the studied parameters. Furthermore, another set of simulations were performed, based on an array of disc-shaped nanostructures with intrinsic shape anisotropy, which revealed a dependance of the charecteristic fields with the direction of the applied external field. It wasalso inferred that the discs in the extremities (with less close neighbours) were the first ones to nucleate (or annihilate) the vortex state. This difference was attributed to the smaller overal interaction of the neighbouring discs.

References

[1] S. Bohlens, B. Krüger, A. Drews, M. Bolte, G. Meier, D. Pfannkuche, Applied Physics Letters **93**, 142508 (2008), `https://doi.org/10.1063/1.2998584`

[2] Y. Cheng, M.E. Muroski, D.C. Petit, R. Mansell, T. Vemulkar, R.A. Morshed, Y. Han, I.V. Balyasnikova, C.M. Horbinski, X. Huang et al., Journal of Controlled Release **223**, 75 (2016), `15334406`

[3] A. Wachowiak, J. Wiebe, M. Bode, O. Pietzsch, M. Morgenstern, R. Wiesendanger, Science **298**, 577 (2002)

[4] T. Shinjo, T. Okuno, R. Hassdorf, †.K. Shigeto, T. Ono, Science **289**, 930 (2000), `http://science.sciencemag.org/content/289/5481/930.full.pdf`

[5] K.Y. Guslienko, V. Novosad, Y. Otani, H. Shima, K. Fukamichi, Applied Physics Letters **78**, 3848 (2001)

[6] A. Vansteenkiste, J. Leliaert, M. Dvornik, M. Helsen, F. Garcia-Sanchez, B.V. Waeyenberge, AIP Advances **4**, 107133 (2014)

[7] G.S. Abo, Y.K. Hong, J. Park, J. Lee, W. Lee, B.C. Choi, IEEE Transactions on Magnetics **49**, 4937 (2013)

[8] O.V. Salata, Journal of Nanobiotechnology **2**, 1 (2004), NIHMS150003
[9] V. Novosad, K.Y. Guslienko, H. Shima, Y. Otani, S.G. Kim, K. Fukamichi, N. Kikuchi, O. Kitakami, Y. Shimada, Physical Review B **65**, 060402 (2002)
[10] K.Y. Guslienko, V. Novosad, Y. Otani, H. Shima, K. Fukamichi, Physical Review B - Condensed Matter and Materials Physics **65**, 244141 (2002)
[11] X. Zhu, V. Metlushko, B. Ilic, P. Grutter, IEEE Transactions on Magnetics **39**, 2744 (2003)
[12] N. Nishimura, T. Hirai, A. Koganei, T. Ikeda, K. Okano, Y. Sekiguchi, Y. Osada, Journal of Applied Physics **91**, 5246 (2002)
[13] R.P. Cowburn, Science **287**, 1466 (2000)
[14] J. d'Albuquerque e Castro, D. Altbir, J.C. Retamal, P. Vargas, Physical Review Letters **88**, 2372021 (2002)
[15] B. Mora, A. Perez-Valle, C. Redondo, M.D. Boyano, R. Morales, ACS Applied Materials & Interfaces **10**, 8165 (2018)
[16] A. Fernandez, C.J. Cerjan, Journal of Applied Physics **87**, 1395 (2000)

Magnetostriction in Amorphous Co66Fe34 Microcantilevers Fabricated with Hydrogenated Amorphous Silicon

B. M. Silveira[1,2], *J. H. Belo*[1], *R. Pinto*[3], *J. A. Silva*[1], *T. D. Ferreira*[2], *A. L. Pires*[1], *V. Chu*[3], *J. P. Conde*[3,4], *O. Frazão*[2], *and A. M. Pereira*[1]*

[1]IFIMUP-IN – Institute for Nanoscience and Nanotechnology, Rua do Campo Alegre, 4169-007, Porto, Portugal

[2]INESC-TEC – Technology and Science, Rua do Campo Alegre, 4169-007, Porto, Portugal

[3]INESC-MN – Microsystems and Nanotechnologies, Rua Alves Redol, 1000-029 Lisboa, Portugal

[4]Department of Bioengineering, Instituto Superior Técnico, Universidade de Lisboa, 1049-001 Lisboa, Portugal

Abstract. To study the magnetostriction of $Co_{66}Fe_{34}$ thin films, amorphous silicon microcantilevers were prepared by surface micromachining, and the 136 nm-thick magnetostrictive film was deposited by electron beam physical vapor deposition and patterned on top of the microcantilever structure. The magnetostriction of the $Co_{66}Fe_{34}$ films was confirmed by measuring the deflection of the cantilevers under a varying magnetic field, reaching displacements up to 8 nm. The configuration was simulated using COMSOL software, yielding a similar deflection behavior as a function of the magnetic field, with a film with a magnetostrictive coefficient of $\lambda_S \sim 55$ p.p.m. The experimental configuration uses a laser and a position sensitive detector to measure the displacement, based on an optical lever configuration, and a piezoelectric stage to calibrate the system.

Keywords: Magnetostriction, Thin films, Micromechanical & nanomechanical oscillators, Functional Materials, Evaporation, Bending

1 Introduction

Internet of things, the so-called IoT, and the Internet of Energy, the so-called IoE, are emerging in our daily lives through the development of new applications towards Smart Cities [1,2]. Such unscalable growth is fostering the market through the introduction of new sensor-enabled mobile and web applications almost every day [1]. A significant social impact is imminent, namely by allowing the emergence of IoT ecosystems supported by open technologies and platforms [1,2]. Although this tremendous evolution of IoT is being backed by new software, advances at a hardware level are still required. Thus, the demand for innovative multifunctional sensors in the near future is urgent, specifically sensors with new functionalities, and higher sensitivity [3].

The field of magnetic sensors/actuators is one of the fields that is being boosted in this new era and, although being studied for many years now, the combination of the magnetostrictive effect with micro-electro-mechanical systems (MEMS) is still at an early stage. MEMS are being applied in many research

[1] *Corresponding author: ampereira@fc.up.pt
Phone: + 351 220402369
Fax: + 351 220402369

areas such as magnetic field sensing (high magnetic fields or ultra-low magnetic fields) [4,5], actuator systems for wireless devices [6], or biological applications, namely in platforms for mass and force detection [7]. In the '90s, Koch and co-authors [8] demonstrated the use of a cantilever beam technique as a potential magnetometer for the quantitative measurement of magnetization, magnetostriction, and magnetocrystalline anisotropy. It took one decade to demonstrate that an optical method could be implemented to detect the bending of a cantilever and that it could be achieved by depositing a magnetostrictive material on the cantilever [9]. Several efforts to combine magnetic systems and motion of miniaturized vibrating mechanical structures have been reported. Herein, research on magnetic material characteristics [10,11] and sensitive detection of an external magnetic field [12,13] should be highlighted. More recently, Park and co-authors [14] showed that by using Ni thin films, it is possible to tailor the resonant behavior of miniaturized mechanical systems by controlling the surface stress originated by the magnetostrictive film. The authors reported an effective magnetostriction of 260 p.p.m. at a low-saturation field of $\approx$ 10 mT for an atomic composition close to Co:Fe = 66:34. In the literature, the atomic composition of Co:Fe = 66:34 is typically associated to be responsible for the highest magnetostrictive effect [15]. Electroplating [16] and DC sputtering [17] have been reported for the production of such thin films. Despite the achieved breakthrough, only a few works on thin film preparation of this promising intermetallic compound were subsequently performed.

Most magnetostrictive applications are based on polycrystalline or single crystal Si substrates [14,16,18–20] and, to the best of our knowledge, no studies on magnetostrictive cantilevers using hydrogenated amorphous silicon (a-Si:H) as structural layer, which presents high potential for MEMS applications due to the low temperature process and reduced residual stress [21], have been published yet.

In the present work, the production of a magnetostrictive microcantilever based on a-Si:H and amorphous $Co_{66}Fe_{34}$ thin films are addressed and tested. Numerical simulations to support the results are performed, using a finite elements method.

2 Experimental Details

The surface and chemical composition of the produced films and MEMS devices were evaluated by Scanning Electron Microscopy (SEM) using Phillips-FEI/Quanta 400 FEG, from Materials Center of the University of Porto (CEMUP), with a coupled Energy Dispersive Spectroscopy (EDS). X-ray diffraction (XRD) measurements were carried out, at room temperature (RT), using a Rigaku Smartlab X-ray diffractometer in the Bragg-Brentano configuration, operating at 9 kW, with the K_α line of Copper corresponding to a wavelength of 1.5418Å. Magnetic properties of the bulk sample were obtained by measuring the M–H curve with a Vibrating Sample Magnetometer (VSM), KLA Tencor EV7, operating at RT and with an applied magnetic field varying in the range between $\pm$ 0.8 T.

To measure the deflection of the free end of the microcantilevers as a function of the applied magnetic field, the microcantilevers were placed between two Helmholtz coils using a commonly implemented optical lever setup, as illustrated in Fig. 1.

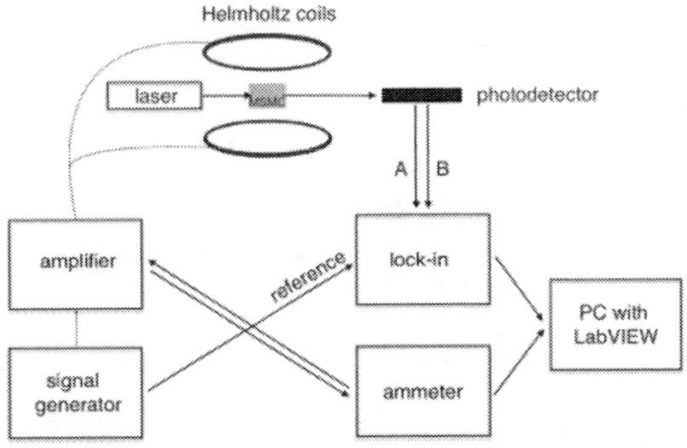

Fig. 1. Schematic representation of the measurement setup. An AC magnetic field is induced in the Helmholtz coils by amplifying a sinusoidal voltage signal from the signal generator. The varying magnetic field causes the microcantilevers to bend and the deflection is measured using a PSD connected to a lock-in, which in turn is connected to a computer.

The microcantilevers were placed so that the magnetic field was applied parallel to their length (longitudinal). A BOP 50-4D Kepco was used to amplify a sinusoidal signal from a signal generator (Agilent 33220A), which was injected into the coils, producing an AC magnetic field. A collimated laser source (658 nm; 10 μW) was used to shine light on a microcantilever tip, and an Eltec 404-4 position sensitive detector (PSD) was used to collect the reflected signal [15,22]. The laser beam was pointed at one microcantilever at a time. The varying magnetic field leads to the change in volume of the magnetostrictive thin film, resulting in the deflection of the microcantilevers. The amplitude of the measured signal, R, and the phase between the signal measured in the photodetector and the reference signal from the signal generator, ϕ, were obtained by a lock-in (Stanford Research Systems SR830), which was connected to the photodetector. Since the AC current in the coils induces a vibrating movement on the microcantilevers, the lock-in was set to be driven at twice the reference frequency to measure the second harmonic. The integration time of the signal ranged from 1 s to 30 s. A computer with LabVIEW was used for data acquisition. To convert the values of R into displacement, a PSI-5A4E piezoelectric (PZT) sheet was placed beneath the microcantilevers and the PZT-induced motion of the non-actuated cantilever was individually calibrated. The calibration was performed by applying an increasing voltage between the upper and bottom surfaces of the PZT, and measuring the amplitude R of the first harmonic in the lock-in. The values for the applied magnetic field were measured with a Hall probe.

3 Microfabrication of Magnetostrictive Cantilevers

3.1 Magnetostrictive $Co_{66}Fe_{34}$ Thin Film Deposition

The $Co_{66}Fe_{34}$ target material was produced by arc melting using as starting reagents Alfa Aesar Co pieces with 99.9+% purity and Fe pieces with 99.97+% purity. After several high-vacuum purges, an Argon working pressure of 700 mbar was used for the arc-melting of the starting reagents. Electron Beam Physical Vapor Deposition (EB-PVD) from Edwards Auto 306 system was used to produce the $Co_{66}Fe_{34}$ films. A 10 mA current and a 4 kV voltage were applied on a Tungsten filament. The kinetic energy of the generated electron beam is converted into thermal energy that heats the target located underneath. The procedure adopted in this work was of small deposition time intervals of 3 minutes, in a total of 72 minutes, with 5-minute intervals between each deposition to avoid overheating. Magnetostrictive thin films were achieved with thicknesses up to 136 ± 5 nm.

3.2 Fabrication of Microcantilevers with Magnetostrictive Film

The microcantilevers were fabricated by surface-micromachining using a 4-mask process. The fabrication started with the cleaning of the substrate, a 0.7 mm-thick glass (Corning 1737), followed by the deposition of a 1.02 μm-thick Al sacrificial layer using a Nordiko 7000 DC magnetron sputtering system. Optical lithography was performed by direct write laser (DWL) using a positive photoresist (PFR 7790G-27cP, from JSR Micro), followed by wet etching (Gravure Aluminium Etchant Micropur MOS from Technic, France). Then, the structural layer of the microcantilevers, an 850 nm-thick layer of n-doped hydrogenated amorphous silicon (n+a-Si:H), was deposited by plasma enhanced chemical vapor deposition, defined using DWL and patterned by reactive ion etching [23]. Another lithography process was performed on top of the cantilevers to define the region where the magnetostrictive film was to be deposited.

The $Co_{66}Fe_{34}$ film was deposited by EBPVD using intermittent deposition time intervals to avoid overheating. This methodology was chosen based on previous and already optimized tests. This step was followed by liftoff in Microstrip 3001 (Fujifilm). In order to protect the magnetostrictive film during the microcantilever release, the last lithography step was performed in order to define photoresist rectangles on top of the $Co_{66}Fe_{34}$ regions. Finally, the release of the microcantilevers was achieved using Al etchant in order to remove the Al sacrificial layer. After release, the dies were sequentially washed in water (to remove the Al etchant), acetone (to remove the photoresist), isopropanol and n-hexane (solvents with decreasing surface tension, to avoid stiction), and let dry at ambient temperature.

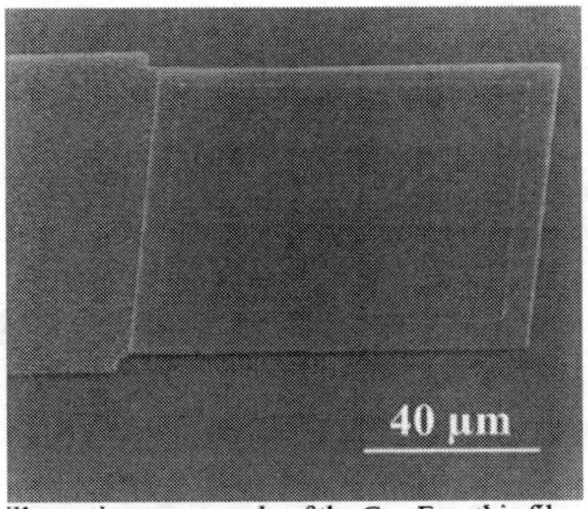

Fig. 2. Top-view SEM micrograph illustrating an example of the $Co_{66}Fe_{34}$ thin film deposited on top of the micrometric cantilever throughout an area smaller than the cantilever area.

4 Results and Discussion

4.1 Magnetostrictive Microcantilevers

A proof-of-concept experiment was undertaken to demonstrate the magnetostrictive behavior of the $Co_{66}Fe_{34}$ thin films, by depositing a layer of this material over a microstructure (Fig. 2a) and b)).

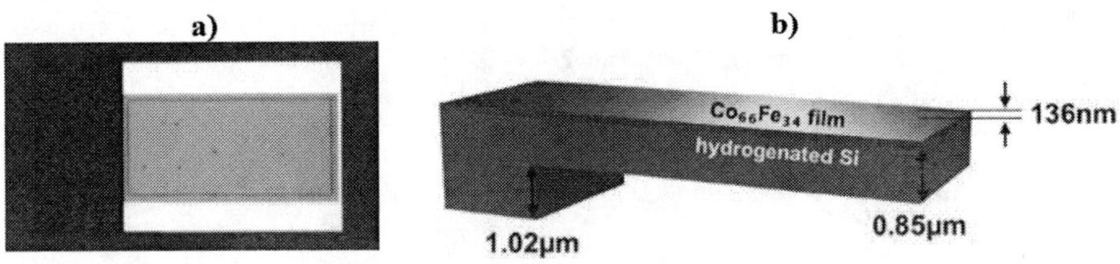

Fig. 2. Magnetostrictive microcantilever (MSMC): a) top view optical microscopy image and b) schematic illustration of the lateral view.

In the first attempt to integrate magnetostrictive films with thin-film silicon cantilevers, the magnetostrictive thin films were deposited by EBPVD using a continuous 80 min deposition. However, as observed in Fig. 3, the high temperature reached by the magnetostrictive material during its deposition degraded the microcantilever. This degradation can be associated with the higher thermal expansion coefficient of Al in comparison to those of silica and $Co_{66}Fe_{34}$.

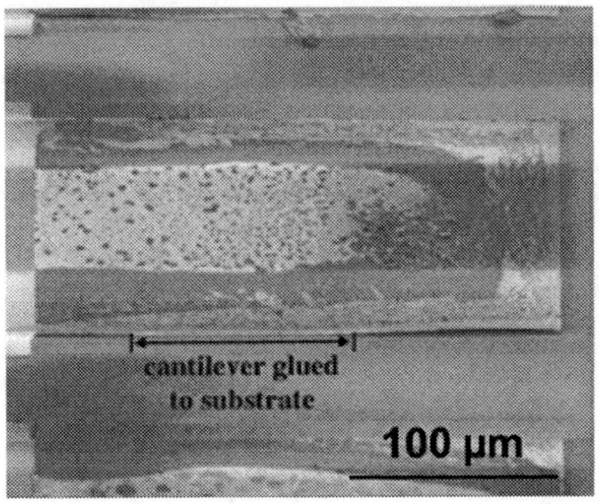

Fig. 3. Top-view SEM imaging shows a damaged microcantilever in contact with the substrate (stiction).

To overcome the destruction of the microcantilevers caused by overheating during the deposition of the magnetostrictive material, the process was conducted with intermittent deposition times followed by a waiting period. Thus, the thin films were deposited using small time intervals of 3 minutes, in a total of 72 minutes, with 5-minute intervals between each deposition. Using a *Dektak XT* profilometer from *Bruker*, a thin film thickness of approximately 136 ± 5 nm was achieved.

In Figure 4, the XRD pattern of the MSMC is depicted.

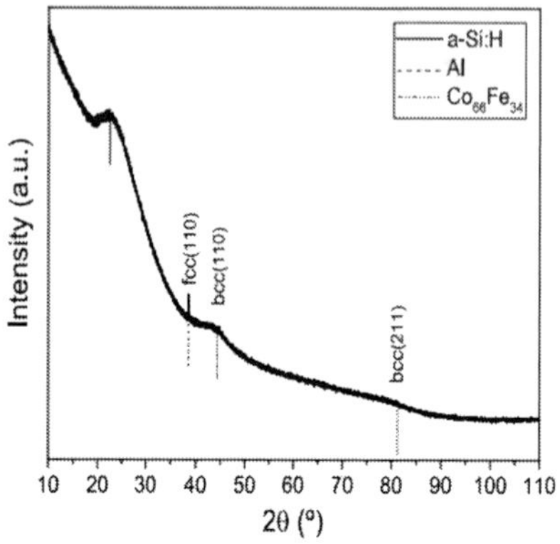

Fig. 4. The X-ray diffractogram for the MSMC measured at RT showing substantially broad peaks, associated with a non-crystalline $Co_{66}Fe_{34}$ layer over the microcantilevers (linear intensity for better reading of the plot).

The XRD was performed on the MSMCs before removing the photoresist and the Al sacrificial layer. The first broad band at $2\theta \approx 23^{\circ} - 26^{\circ}$ is indexed to amorphous/nanocrystalline n^{+}a-Si:H layer [24]. Additional Bragg reflections are present at $2\theta \approx 38^{\circ}$ corresponding to the Al fcc(110) phase [25], and at $2\theta \approx 45^{\circ}$ and $2\theta \approx 83^{\circ}$ corresponding to the bcc(110) and bcc(211) planes of the phase, respectively, indicating the presence of the Co-Fe compound [26]. Nonetheless, the last referenced peaks are broad and weak, which is associated with a lack of a long-range crystalline organization characteristic of the amorphous/nanocrystalline structure.

4.2 Displacement Measurements

As proof of concept, the presence of a magnetostrictive effect was confirmed by magnetically actuating the MSMC movement using the configuration illustrated in Fig. 1.

Fig. 5 depicts the displacement for the microcantilever M_1 (100 µm × 60 µm) as a function of the amplitude of AC magnetic field. It is possible to observe a parabolic increase of displacement with the applied magnetic field. To test the repeatability of the deflection method using this characterization setup, two measurements were carried out using M_1 under the same conditions and for the same frequency. The observed shift on the displacement curves its minimal, which indicates that the microcantilever was not critically damaged for such low fields and that it returns to its initial geometric configuration after the magnetic field is removed.

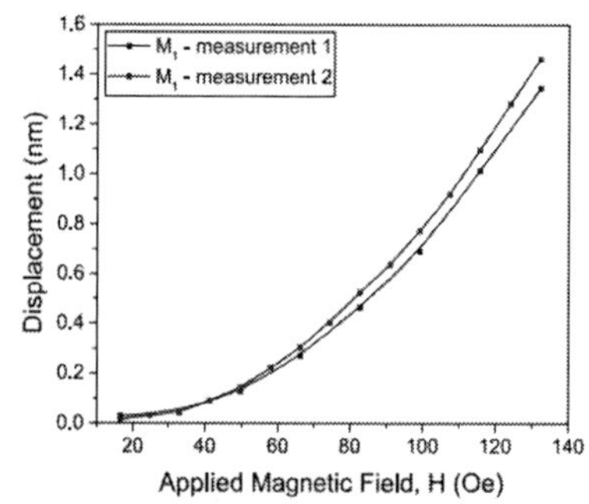

Fig. 5. Displacement as a function of the applied magnetic field for microcantilever M_1 was obtained using the configuration schematized in Fig. 1. Nanometric displacements were measured over the magnetic field range considered, and the results show acceptable repeatability.

In order to have an estimation of the magnetostrictive coefficient of the $Co_{66}Fe_{34}$ thin film, numerical simulations were performed with the COMSOL Multiphysics software. By combining the Structural Mechanics and the AC/DC modules of the software, the cantilever M_1 was simulated inside a box of air. The Young's modulus of the n^+a-Si:H and $Co_{66}Fe_{34}$ were set to 80 GPa [27] and 140 GPa [15], respectively. Since it was not possible to measure the magnetic properties of the films deposited on the microcantilevers (as it would critically damage the microstructure), the magnetic behavior of the ferromagnetic film was considered to be similar to that of the amorphous CoFeB film reported by Jen and Yao [28]. An isotropic model for the magnetostrictive effect was used, and the coefficient was varied to fit the experimental results.

The best resemble simulation is depicted in Fig. 6 and the corresponding coefficient obtained was $\lambda_S = 55 \times 10^{-6}$ (55 p.p.m.). The achieved coefficient is in good agreement with the literature. For instance, Cooke [29] and Hunter [15] groups reported about 80 p.p.m. in as-deposited thin films of $Co_{66}Fe_{34}$. In comparison with annealed $Co_{66}Fe_{34}$ thin films coefficient, the obtained value is low, however this was expected for the nanocrystalline/amorphous nature of the thin film here reported.

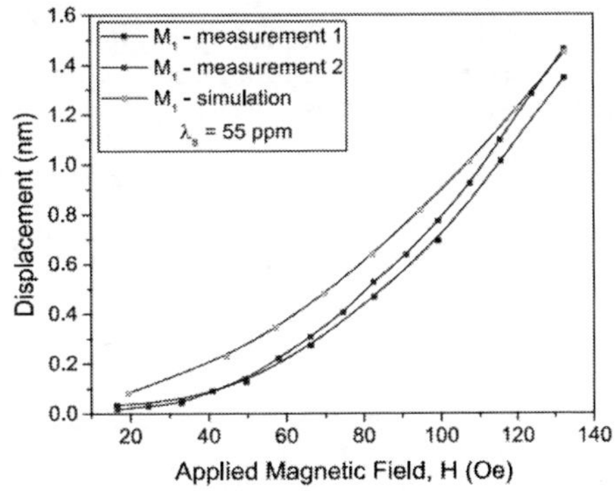

Fig. 6. Results from the simulation performed with COMSOL Multiphysics software and comparison with experimental data. This study allows us to conclude that the magnetostrictive coefficient of the $Co_{66}Fe_{34}$ film deposited on n+a-Si:H is around 55×10^{-6} (55 p.p.m.).

5 Conclusions

A microcantilever structure was developed using micro-fabrication techniques to verify magnetostriction in amorphous $Co_{66}Fe_{34}$ thin films. The optical characterization system based on the reflection of laser light on the surface of the MSMC allowed for the measurement of microcantilever nano-displacements, for applied magnetic fields ranging from [0;140] Oe, verifying magnetostriction on amorphous $Co_{66}Fe_{34}$ films. Numerical simulations performed with COMSOL Multiphysics software have allowed an estimation of the magnetostrictive coefficient of the MSMC M_1 up to 55 p.p.m., which is in agreement with the literature. In the future, an additional buffer layer needs to be implemented to minimize the influence of the thermal expansion.

Acknowledgment

This work was financially supported by Fundação para a Ciência e a Tecnologia (FCT)/MEC and FEDER under Program PT2020 through the projects UID/NAN/50024/2019, and NORTE-01-0145-FEDER-022096 from NECL. A.M.P. and J.A.S. acknowledge ANI and FCT through the project POCI-01-0247-FEDER-033783. J.H.B acknowledges the Grant SFRH/BD/88440/2012, the project PTDC/FIS-MAC/31302/2017 and his contract DL57/2016 reference SFRH-BPD-87430/2012. R.M.R.P. (PD/BD/113960/2015) acknowledges FCT for a PhD Grant, through the research project uMEMS (PTDC/CTM-NAN/5052/2014), and for funding through the Research Unit IN–Institute of Nanoscience and Nanotechnology (UID/NAN/50024/2013).

References

[1] D. Miorandi, S. Sicari, F. De Pellegrini, I. Chlamtac, I Ad Hoc Networks. 10 (2012) 1497–1516.
[2] T. Islam, S.C. Mukhopadhyay, N.K. Suryadevara, IEEE Sens. J. 17 (2017) 577–584.

[3] A. Rayes, S. Salam, Springer International Publishing, Cham, 2019.
[4] F. Keplinger, S. Kvasnica, H. Hauser, R. Grossinger, IEEE Trans. Magn. 39 (2003) 3304–3306.
[5] A. Guedes, G. Jaramillo, C. Buffa, G. Vigevani, S. Cardoso, D.C. Leitao, P.P. Freitas, D.A. Horsley, IEEE Trans. Magn. 48 (2012) 4115–4118.
[6] B.A. Cetiner, H. Jafarkhani, Jiang-Yuan Qian, Hui Jae Yoo, A. Grau, F. De Flaviis, IEEE Commun. Mag. 42 (2004) 62–70.
[7] X. Li, D.W. Lee, Meas. Sci. Technol. 23 (2012) 029501.
[8] M. Weber, R. Koch, K.H. Rieder, Phys. Rev. Lett. 73 (1994) 1166–1169.
[9] D. Sander, J. Kirschner, Appl. Phys. A. 87 (2007) 419–425.
[10] G. Wu, R. Zhang, X. Li, N. Zhang, J. Appl. Phys. 110 (2011) 124103.
[11] X.W. Dong, Y.J. Wu, J.G. Wan, T. Wei, Z.H. Zhang, S. Chen, H. Yu, J.M. Liu, J. Phys. D. Appl. Phys. 41 (2008) 035003.
[12] T. Nan, Y. Hui, M. Rinaldi, N.X. Sun, Sci. Rep. 3 (2013) 1985.
[13] B. Gojdka, R. Jahns, K. Meurisch, H. Greve, R. Adelung, E. Quandt, R. Knöchel, F. Faupel, Appl. Phys. Lett. 99 (2011) 223502.
[14] Y. Park, E. Lee, T. Kouh, Micromachines. 8 (2017) 109.
[15] D. Hunter, W. Osborn, K. Wang, N. Kazantseva, J. Hattrick-Simpers, R. Suchoski, R. Takahashi, M.L. Young, A. Mehta, L. Bendersky, S.E. Lofland, M. Wuttig, I. Takeuchi, Nat. Commun. 2 (2011) 518.
[16] B. Özkale, N. Shamsudhin, T. Bugmann, B.J. Nelson, S. Pané, Electrochem. Commun. 76 (2017) 15–19.
[17] J. Zhou, S. Chen, W. Lin, Q. Qin, L. Liu, S. He, J. Chen, J. Magn. Magn. Mater. 441 (2017) 264–270.
[18] L. Fu, S. Li, K. Zhang, I.H. Chen, V. Petrenko, Z. Cheng, Sensors. 7 (2007) 2929–2941.
[19] G. Engdahl, E. Quandt, Handbook of Giant Magnetostrictive Materials, 2000.
[20] Y.Y. Kim, Y.E. Kwon, Ultrasonics. 62 (2015) 3–19.
[21] J. Mouro, A. Gualdino, V. Chu, J.P. Conde, J. Appl. Phys. 114 (2013) 184905.
[22] G. Meyer, N.M. Amer, Appl. Phys. Lett. 53 (1988) 1045–1047.
[23] J. Mouro, V. Chu, J.P. Conde, J. Appl. Phys. 119 (2016) 154501.
[24] T.V. Torchynska, Phys. E Low-Dimensional Syst. Nanostructures. 44 (2011) 56–61.
[25] S.W. Lin, J.Y. Wu, S.D. Lin, M.C. Lo, M.H. Lin, C.T. Liang, Jpn. J. Appl. Phys. 52 (2013) 045801.
[26] T. Nakajima, T. Takeuchi, I. Yuito, K. Kato, M. Saito, K. Abe, T. Sasaki, T. Sekiguchi, S. Yamaura, Mater. Trans. 55 (2014) 556–560.
[27] L.B. Freund, S. Suresh, Thin film materials, 2004.
[28] S.U. Jen, Y.D. Yao, Y.T. Chen, J.M. Wu, C.C. Lee, T.L. Tsai, Y.C. Chang, J. Appl. Phys. 99 (2006) 053701.
[29] M.D. Cooke, M.R.J. Gibbs, R.F. Pettifer, J. Magn. Magn. Mater. 237 (2001) 175–180.

CdS versus ZnSnO buffer layers for a CIGS solar cell: a depth-resolved analysis using the muon probe

Eduardo Ribeiro[1], *Helena* V. Alberto[1,*], *Rui* C. Vilão[1], *João* M. Gil[1], *Alois* Weidinger[2], *Pedro* M. P. Salomé[3,4], *Thomas* Prokscha[5], *Andreas* Suter[5], and *Zaher* Salman[5]

[1]CFisUC, Department of Physics, University of Coimbra, R. Larga, P-3004-516 Coimbra, Portugal
[2]Helmholtz-Zentrum Berlin für Materialien und Energie, 14109 Berlin, Germany
[3]International Iberian Nanotechnology Laboratory, 4715-330 Braga, Portugal
[4]Department of Physics, University of Aveiro, 3810-193 Aveiro, Portugal
[5]Laboratory for Muon Spin Spectroscopy, Paul Scherrer Institut, CH-5232 Villigen PSI, Switzerland

Abstract. The influence of a buffer layer in the surface of a Cu(In,Ga)Se_2 (CIGS) solar cell material is studied using implanted positive muons as a probe. A depth resolved analysis of the muon data suggests that both CdS and ZnSnO reduce the width of a defect layer present at the CIGS surface to about half its original value. Additionaly, CdS is able to reduce the intensity of the disturbance in the defected region, possibly due to a surface reconstrution in CIGS.

1 Introduction

Thin film solar cells using Cu(In,Ga)Se_2 (CIGS) as the p-type absorber achieved recently 23.3% solar cell efficiency[1], the highest value among multi-crystalline solar cells. CdS has been the most widely used n-type buffer layer for CIGS solar cell, but has the important disadvantage of containing a toxic element, the cadmium. The material ZnSnO was shown to constitute a good alternative as a Cd-free buffer layer in CIGS, exhibiting a similar performance to CdS [2, 3].

A previous study [4] using implanted positive muons with energies in the range 2-24 keV has shown that the formation of the final bound configuration of the muon is sensitive to local barriers and that the spatial extent of the decrease in formation probability of the muon state can be used as a measure of the width of a defective region ocurring at the p-n interface.

Recently, a new method of analysis of the energy-dependent muon data was developed, allowing a depth-resolved analysis of all the experimental parameters associated with the muon signal [5]. This paper presents the results obtained for low-energy muon data for a CIGS film and two junctions, CdS/CIGS and ZnSnO/CIGS reanalysed using this new method. It is shown that this depth-resolved analysis provides new insights on the microscopic processes ocurring during the formation of the buffer/absorber interface in a CIGS based solar cell.

2 Experimental Details and Data Analysis

Two different heterostructures, CdS/CIGS and ZnSnO/CIGS, were produced in the Ångström Solar Center, Uppsala University, Sweden, from one single CIGS film, using a test pilot

*e-mail: lena@uc.pt

tool for module production. Extensive details on the sample preparation were reported in a previous publication [6]. The thickness of the n-type layer was 50 nm for CdS and 20 nm for ZnSnO. The single CIGS film was obtained by removing the CdS layer from the CdS/CIGS sample by etching with a solution 10% (V/V) of HCl in H_2O, dried with a N_2 blow drier and immediately moved into a N_2 atmosphere to avoid surface oxygenation. It was mounted under continuous N_2 flow in a sample holder and measured in vacuum.

The muon spin spectroscopy (μSR) measurements were performed at the μE4 beam line [7] of the Swiss Muon Source, Paul Scherrer Institut, Switzerland, using the low energy muons (LEM) instrument. Positive muons were implanted in the samples in the presence of an external magnetic field B = 10 mT, using transverse field (TF) geometry at a fixed temperature of T=50 K and variable implantation energy, in order to perform depth dependent studies, as illustrated in Fig.1.

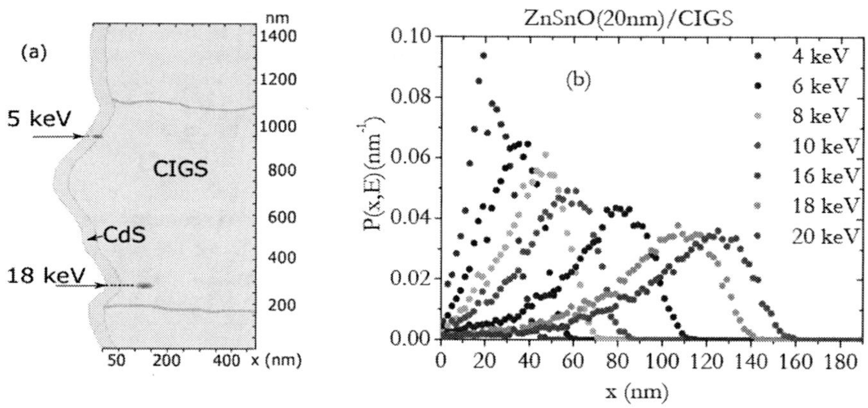

Figure 1. (a): Illustration of the change in implantation depth for two different muon implantation energies, in a CdS/CIGS heterostructure. The drawing is based on a Transmission Electron Microscopy picture. **(b):** Monte Carlo simulations of the probability per unit lenght, $P(x, E)$ that a muon implanted with energy E stops at a depth x in a ZnSnO(20nm)/CIGS junction. The different curves correspond to different muon implantation energies.

The time dependent μSR data corresponds to muons forming a diamagnetic state and was fitted to a Gaussian-damped oscillation at the muon Larmor frequency. The diamagnetic signal amplitude A_{dia} was converted to the corresponding diamagnetic fraction, f_{dia}, using a silver calibration and a sample size correction. The diamagnetic fraction was found to be smaller than unity for most of the measurements. This indicates that the muons experience a strong interaction during their implantation stage, reducing the formation probability of the final bound configuration [4, 8–13].

From the fitting of the muon time spectra, the diamagnetic fraction and the Gaussian relaxation rate were obtained as a function of the muon implantation energy, $f_{\text{dia}}(E)$ and $\sigma(E)$. Figure 1(b), illustrates the muon stopping distribution in a ZnSnO/CIGS sample as a function of sample depth, x, as obtained by Monte Carlo simulation, using the TRIM.SP code [14, 15]. Since the muon stopping region extends over a region of several tens of nanometers, the energy dependence of the relevant μSR parameters cannot be interpreted directly as a depth dependence. Therefore, depth dependent functions were inferred using the method described in reference [5] : the functions $f_{\text{dia}}(x)$ and $\sigma(x)$ were assumed to have a step-like shape with adjustable parameters; $f_{\text{dia}}(E)$ and $\sigma(E)$ functions were then calculated by weighting $f_{\text{dia}}(x)$ and $\sigma(x)$ with the corresponding muon stopping probabilities, obtained by

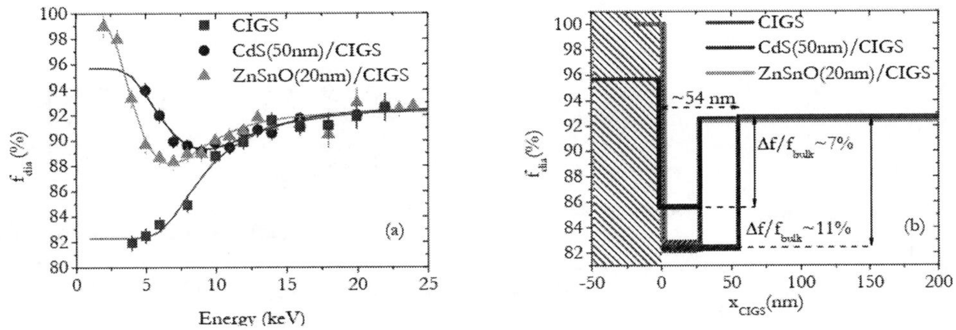

Figure 2. (a) : Diamagnetic fraction as a function of muon implantation energy, $f_{dia}(E)$, at 50 K for CdS/CIGS, ZnSnO/CIGS and CIGS samples. The symbols are the experimental data points. The full curves are the predicted behaviour of the parameter assuming a depth dependence as shown in the corresponding graph at right. **(b)**: Diamagnetic fraction as a function of the distance to the surface of CIGS, $f_{dia}(x_{CIGS})$. The thickness of the lines indicate the uncertainties in the parameters of $f_{dia}(x_{CIGS})$, namely the position and depth of the wells.

numerical integration of the muon stopping probability per unit length, $P(x, E)$ (see Fig.1(b)); the experimental data points were then fitted to the calculated $f_{dia}(E)$ and $\sigma(E)$, in order to optimize the parameters for $f_{dia}(x)$ and $\sigma(x)$. The thickness of the lines in the plots of the functions $f_{dia}(x)$ and $\sigma(x)$ (see Fig.2(b) and Fig.3(b)) are a measure of the uncertainties in the parameters, as obtained in this fitting procedure.

3 Results and Discussion

Figure 2 shows the diamagnetic fraction as a function of muon implantation energy for the three different samples, CdS/CIGS, ZnSnO/CIGS and the CIGS film at 50 K and under an external magnetic field B = 10 mT, together with the corresponding functions $f_{dia}(x_{CIGS})$, where x_{CIGS} is defined as the distance to the CIGS surface.

It is evident from Fig. 2(b) that the fraction of the muon diamagnetic state for the bare CIGS film is about 11% smaller at the surface when compared with its bulk value, within a region which is 54(1) nm wide. The diamagnetic fraction in CIGS is a measure of the formation probability of the final bound configuration of the muon in the lattice and is sensitive to local barriers, such as defects, that may hinder its formation. There are numerous reports that the surface of CIGS is a defect-rich region, which has a slightly different composition from the bulk, namely a smaller Cu content [16–18]. We attribute the dip in the diamagnetic fraction as due to this surface defect layer (SDL) and the observed spatial extent of the dip (~ 54 nm) as a measure of the SDL thickness in the bare CIGS film.

Figure 2(b) also shows that when a top layer of ZnSnO or CdS is added, the width of the dip at the CIGS surface is reduced to around half (25 to 28 nm); in both samples, the recovery of the diamagnetic fraction to the bulk value clearly occurs in the inner part of CIGS. A possible explanation is that this reduction is due to the built-in electric field at the p-n interface. It has been suggested that the Cu-depletion at the CIGS surface is due to Cu electro-diffusion due to positive charges at CIGS surface [18]. The formation of a depleted, negatively charged region in CIGS when CdS or ZnSnO is added may cause a diffusion of Cu from the bulk towards the surface, causing a healing in the inner part of the defect layer.

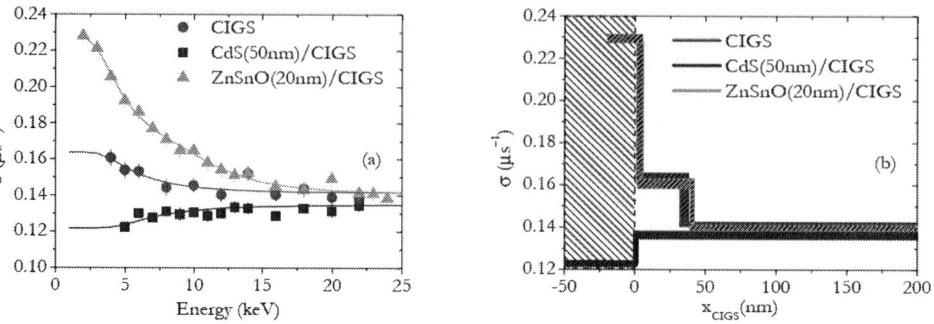

Figure 3. **(a)** : Muon spin relaxation as a function of muon implantation energy, $\sigma(E)$, at 50 K for CdS/CIGS, ZnSnO/CIGS and CIGS samples. The full curves are the predicted behaviour of the parameter assuming a depth dependence as shown in the corresponding graph at right. **(b)**: muon spin relaxation as a function of the distance to the CIGS layer, $\sigma(x_{CIGS})$ for the three samples. The thickness of the lines is a measure of the uncertainties in the parameters of the $\sigma(x_{CIGS})$ function.

Additionally, Fig. 2(b) shows that the depth of the diamagnetic dip in the CIGS region is also reduced for the CdS/CIGS sample (from ~ 11% in CIGS film to ~ 7% in CdS/CIGS) but a similar reduction is not observed for the ZnSnO/CIGS sample. Therefore CdS causes changes both at the surface and inner part of the SDL in CIGS, whereas ZnSnO only affects the SDL inner region.

Figure3 shows that in the CIGS film there is a surface region with a spatial extent of ~ 35 nm, where the muon spin relaxation is ~ 14% larger than in bulk. The increase of the muon spin relaxation at the CIGS surface suggests a change in the muon environment in its final state configuration, caused either by structural disorder, defects, or electron capture. This region at the CIGS surface where the muon suffers a stronger relaxation is still present when ZnSnO is added on top, but disappears if the top layer is CdS. This suggests again that CdS causes a surface reconstruction of CIGS surface, close to the p-n interface, but that does not occur for ZnSnO. These results are consistent with evidences that when a layer of CdS is added on top of CIGS there is a inter-diffusion of ions at the interface, but no such intermixing is found when the top layer is ZnSnO [3, 19].

4 Conclusions

In conclusion, whereas the diamagnetic muon fraction is sensitive to the barrier for the formation of the final muon state, the muon spin relaxation is sensitive to the muon environment in its final configuration. The combination of the information from both parameters, obtained by a depth-resolved analysis of the muon data, suggests that there is a surface defect layer in CIGS material with a spatial extent of around 54 nm and that the width of this region is reduced to about half its original value when either a CdS or ZnSnO layer is added. However, unlike ZnSnO, CdS is also able to induce a surface reconstrution of CIGS, reducing the intensity of the disturbance at the CIGS surface.

Acknowledgement

This work is based on experiments performed at the Swiss Muon Source (SμS), Paul Scherrer Institute, Villigen, Switzerland. This work was supported with funds from the Portuguese National Budget through Fundação para a Ciência e Tecnologia/MCTES and also from the European Regional Development Fund (ERDF) through the Competitiveness and Internationalisation Operational Programa (PT-COMPETE 2020), under projects UID/FIS/04564/2016 and PTDC/FISMAC/29696/2017.

References

[1] M.A. Green, E.D. Dunlop, D.H. Levi, J. Hohl-Ebinger, M. Yoshita, A.W.Y. Ho-Baillie, Progress in Photovoltaics: Research and Applications **27** (2019)

[2] J. Lindahl, J.T. Wätjen, A. Hultqvist, T. Ericson, M. Edoff, T. Törndahl, Progress in Photovoltaics: Research and Applications **21**, 1588 (2013)

[3] P.M.P. Salomé, R. Ribeiro-Andrade, J.P. Teixeira, J. Keller, T. Törndahl, N. Nicoara, M. Edoff, J.C. González, J.P. Leitão, S. Sadewasser, IEEE Journal of Photovoltaics **7**, 858 (2017)

[4] H.V. Alberto, R.C. Vilão, R.B.L. Vieira, J.M. Gil, A. Weidinger, M.G. Sousa, J.P. Teixeira, A.F. da Cunha, J.P. Leitão, P.M.P. Salomé et al., Phys. Rev. Materials **2**, 025402 (2018)

[5] A.F.A. Simões, H.V. Alberto, R.C. Vilão, J.M. Gil, J.M.V. Cunha, M.A. Curado, P.M.P. Salomé, T. Prokscha, A. Suter, Z. Salman, submitted to Review of Scientific Instruments (2019)

[6] P.M.P. Salomé, J.P. Teixeira, J. Keller, T. Törndahl, S. Sadewasser, J.P. Leitão, IEEE Journal of Photovoltaics **7**, 670 (2017)

[7] T. Prokscha, E. Morenzoni, K. Deiters, F. Foroughi, D. George, R. Kobler, A. Suter, V. Vrankovic, Nuclear Instruments and Methods in Physics Research Section A: Accelerators, Spectrometers, Detectors and Associated Equipment **595**, 317 (2008)

[8] R.C. Vilão, R.B.L. Vieira, H.V. Alberto, J.M. Gil, A. Weidinger, Phys. Rev. B **96**, 195205 (2017)

[9] R.C. Vilão, R.B.L. Vieira, H.V. Alberto, J.M. Gil, A. Weidinger, R.L. Lichti, P.W. Mengyan, B.B. Baker, J.S. Lord, Phys. Rev. B **98**, 115201 (2018)

[10] R.C. Vilão, H.V. Alberto, J.M. Gil, A. Weidinger, Phys. Rev. B **99**, 195206 (2019)

[11] H.V. Alberto, A. Weidinger, R.C. Vilão, J.P. Duarte, J.M. Gil, J.S. Lord, S.F.J. Cox, Phys. Rev. B **81**, 245205 (2010)

[12] H.V. Alberto, R.C. Vilão, J.P. Duarte, J.M. Gil, A. Weidinger, J.S. Lord, S.F.J. Cox, Phys. Rev. B **86**, 035203 (2012)

[13] H.V. Alberto, R.C. Vilão, J.M. Gil, J. Piroto Duarte, R.B.L. Vieira, A. Weidinger, J.P. Leitão, A.F. da Cunha, M. Sousa, J.P. Teixeira et al., Journal of Physics: Conference Series **551**, 012045 (2014)

[14] E. Morenzoni, H. Glückler, T. Prokscha, R. Khasanov, H. Luetkens, M. Birke, E. Forgan, C. Niedermayer, M. Pleines, Nuclear Instruments and Methods in Physics Research Section B: Beam Interactions with Materials and Atoms **192**, 254 (2002)

[15] W. Eckstein, *Computer Simulations of Ion-Solid Interactions* (Springer, Berlin, Heidelberg, New York, 1991)

[16] D. Liao, A. Rockett, Applied Physics Letters **82**, 2829 (2003)

[17] Y. Yan, K.M. Jones, J. Abushama, M. Young, S. Asher, M.M. Al-Jassim, R. Noufi, Applied Physics Letters **81**, 1008 (2002), `https://doi.org/10.1063/1.1498499`

[18] Herberholz, R., Rau, U., Schock, H. W., Haalboom, T., Gödecke, T., Ernst, F., Beilharz, C., Benz, K. W., Cahen, D., Eur. Phys. J. AP **6**, 131 (1999)

[19] P.M.P. Salomé, J. Keller, T. Törndahl, J.P. Teixeira, N. Nicoara, R.R. Andrade, D.G. Stroppa, J.C. González, M. Edoff, J.P. Leitão et al., Solar Energy Materials and Solar Cells **159**, 272 (2017)

Orthorhombic $GdMnO_3$ Epitaxial Thin Film Grown onto SrTiO3 (110)

P. Machado[1], F. G. Figueiras[1], R. Vilarinho[1], J. R. A. Fernandes[2], P. B. Tavares[3], M. Rosário Soares[4], S. Cardoso[5], A. Almeida[1], J. Agostinho Moreira[1]

[1]IFIMUP e Departamento de Física e Astronomia, Faculdade de Ciências da Universidade do Porto. R. Campo Alegre 687, 4169-007 Porto, Portugal
[2]CQVR e Departamento de Física, Universidade de Trás-os-Montes e Alto-Douro, Ap.º 1013, 5001-801 Vila Real, Portugal
[3]CQVR e Departamento de Química, Universidade de Trás-os-Montes e Alto-Douro, Ap.º 1013, 5001-801 Vila Real, Portugal
[4]CICECO & LCA, Universidade de Aveiro, 3810-193 Aveiro, Portugal
[5]INESC-MN, Rua Alves Redol, 9, 1000-029 Lisboa, Portugal

Abstract. $GdMnO_3$ epitaxial thin films were deposited onto (110)-oriented $SrTiO_3$ substrates by RF magnetron sputtering. The structure, microstructure, dielectric and magnetic properties were investigated in detail. The XRD results revealed that the $GdMnO_3$ thin films exhibit an epitaxial strained orthorhombic symmetry and grow preferably in off-plane (001) orientation, wherein the basal lattice parameters are strained by the substrate lattice. A dielectric relaxation process was ascertained, whose activation energy is sensitive to the magnetic phase transitions, occurring at 41 K and at 20 K, respectively. The canted nature of the low temperature magnetic phase is discussed.

1 Introduction

The demand for new materials with higher performance and versatility has been focused on those exhibiting coupling between different degrees of freedom [1,2]. The knowledge concerning the microscopic mechanisms underlying the coupling between orbital, electronic, spin and lattice degrees of freedom, occurring in transition metal oxides with perovskite structure, has been able to advance in the design of new multifunctional materials. In this regard, the rare-earth orthomanganites ($RMnO_3$) with P*nma* symmetry have been deeply studied as their physical properties are very sensitive to structure due to the spin-orbital-lattice coupling [3,4]. It was established, both experimentally and theoretically, that the interesting magnetic and ferroelectric (multiferroic) properties found for some $RMnO_3$ are constrained by the Mn-O-Mn bond angle [4,5]. Changing the Mn-O-Mn bond angle, the orbital overlap between adjacent Mn^{3+} and O^{2-} ions is altered, which consequently modifies the magnetic superexchange integrals between the magnetic Mn^{3+} ions, crucial for the stabilization of the different spin arrangements [5].

$GdMnO_3$ (GMO) stands at the border between the non-multiferroic and multiferroic regions of the phase diagram of $RMnO_3$ [3,5]. On cooling, GMO undergoes a transition into a collinear-sinusoidal incommensurate antiferromagnetic (AFM) at T_N = 43 K and then a transition to a canted A-type AFM phase at $T_c = 22$ K, both associated with the Mn^{3+} spins [3,5]. The long range order of Gd^{3+} magnetic moments emerges below T_N'= 6.5K [3]. Multiferroicity can be achieved below T_C by applying a magnetic-field above 2 T, wherein a magnetically-order ferroelectric phase emerges via magnetoelectric coupling [6]. Spontaneous ferroelectricity can also be obtained by slightly reducing the Mn-O-Mn bond angle via chemical substitution of the Gd^{3+} ions by Tb^{3+}, Y^{3+}, among others [7,8]. The interplay between structure and magnetism can be changed using structural distortions. Here we highlight that ferroelectricity was achieved below 30 and 75 K on orthorhombic GMO thin-films with 10 and 110 nm, deposited by pulsed laser deposition onto (010)-$YAlO_3$ and (001)-$SrTiO_3$ substrates, respectively [9,10]. Recently, the study of epitaxially strained GMO thin-films deposited on (001)-$SrTiO_3$ substrate by RF magnetron sputtering, conveyed an unprecedently reported tetragonal structure at room temperature and ferroelectric below 32 K, up to 35 nm thick films [11]. The emergence of a spontaneous electric polarization in these films reveal the stabilization of a low temperature improper ferroelectric phase, not observed in bulk GMO. In order to go further in the study of the structure-related physical properties of GMO, we present a structural, magnetic and dielectric study of GMO thin-films deposited onto (110)-$SrTiO_3$ substrates through RF magnetron sputtering method. We aim at unravelling the role played by the crystallographic orientation of the substrate onto the film structure and its physical properties. The results obtained will be examined in detail, and compared with those reported for identically deposited GMO thin-films, though grown on (001)-$SrTiO_3$ substrates.

2 Experimental Details

GMO polycrystalline targets were prepared following the procedure described elsewhere [11]. Polished (110)-oriented $SrTiO_3$ substrates (STO(110)) were previously cleaned before deposition in ultrasonic baths and dried using N_2 flow. GMO thin films were deposited onto STO(110) at 800 °C, through RF-magnetron sputtering, working at 13.56 MHz and 80 W RF power. The distance between target and substrate was 130 mm. The deposition rate was 2 nm/min. The pressure inside the chamber was adjusted to 5×10^{-2} mbar, with a gas composition setting a ratio of $Ar:O_2$ of 4:1. After deposition, the films were cooled naturally in a pure oxygen atmosphere of 1 mbar, without any annealing step. The crystallographic structure was measured by high resolution X-ray diffraction at room conditions, using a X'Pert MRD Philips four-circle diffractometer, in a Bragg–Brentano para-focusing configuration, operating with Cu Kα radiation at 30 mA and 40 kV, and analysed through the Le Bail mode. In-plane specific magnetization measurements were carried out using a SQUID magnetometer, with a resolution better than 5×10^{-7} emu. The complex electric permittivity was measured in the 1 kHz - 1 MHz frequency range with an HP4284A impedance analyser, under a 50 V/cm ac electric field [11]. 50 gold interdigital electrodes (IDEs), with 7 mm length, 20 μm width, 20 μm interspacing and depth of about 90% of the thickness of the deposited film, were used for the measurements [11].

3 Results and discussion

Figure 1(a) displays the X-ray Reflectometry diffractogram of GMO thin film onto STO(110) substrate exhibiting low contrast fringes, from which the thickness of 44 ± 4 nm can be estimated. Figure 1(b) presents the conventional X-ray diffraction pattern (red line). The main STO reflections are observed at 2Θ = 32.14°, 67.56°, assigned to (110) and (220) planes, respectively, confirming the (110)-orientation of the substrate and, ascribing a_{STO} = 3.91(0) Å, in good agreement to referenced data [9].

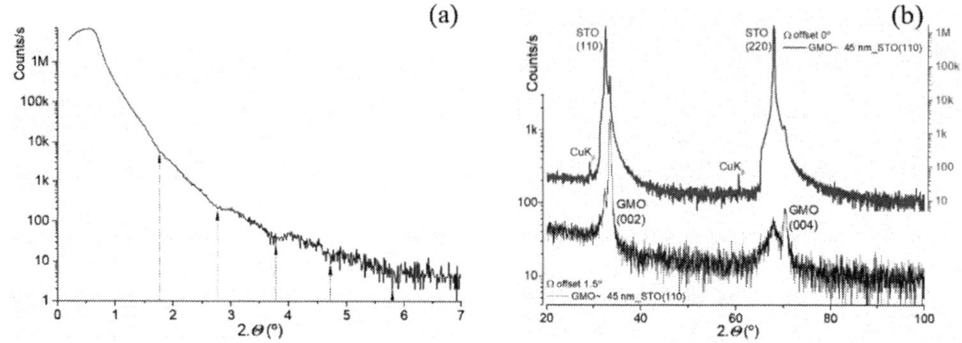

Fig. 1. Diffractograms of the 44 nm GMO/STO(110) thin film (a) XRR with respective fringes signalled by arrows, and (b) XRD pattern (red line) and using a $\Delta\Omega = 1.5°$ offset (blue line).

No evidences of spurious GdO_x or MnO_x phases were found. By adjusting the XRD acquisition parameters with a $\Delta\Omega = 1.5°$ offset (blue line in Figure 1(b)), it allows to focus on the GMO film layer and surpass substrate reflections. The main peaks at $2\Theta = 33.43°$ and $70.24°$ are assigned respectively to the (002) and (004) planes of the GMO phase, evidencing the preferential off-plane growth of GMO oriented along *c*-axis. In addition, it is possible to perceive a slight Bragg peak at $2\Theta = 31.25°$, indexed to the reflection on the (200) planes. This evidences that the films do not grow uniformly and two different GMO layers (hereafter called α- and γ-layers) are present.

Figure 2(a) shows the symmetric XRD space map in the Θ-Ω coordinates, at $2\Theta = 32.43°$ and $\Omega = 16.22°$, centred at the STO(110) reflection peak. This map evidences the presence of 3 reflection peaks emerging from the GMO thin film. The main peak found at $2\Theta = 33.35°$, with $\Omega = 16.72°$ shadows the shape of the STO(110) peak, and shows a long relaxation tail visible up to $2\Theta = 37°$, which can be associated with Laue interference fringes, typically found in a layered topology. This peak is consistent with the reflection on the GMO (002)-planes, confirming the off-plane growth direction along c_γ-axis, and yielding a lattice parameter $c_\gamma = 5.372(4)$ Å, assigned to the γ-layer. Two additional satellite peaks are observed at $2\Theta = 31.24°$, extending orthogonally to each side of STO(110) peak at $\Omega = 18.87°$ and $\Omega = 12.25°$. These peaks are indexed to GMO (200)-plane reflections. Although relatively faint, these two peaks are well-defined ($\Delta 2\Theta \sim \pm 0.5°$, $\Delta\Omega \sim \pm 0.5°$), thus evidencing the presence of a twined crystalline growth along a_α-axis in the off-plane direction, with a calculated lattice parameter $a_\alpha = 5.72(2)$ Å of the α-layer. The low ratio between the I_α/I_γ peaks intensity suggests the α-layer is only a few nanometres thick. The asymmetric space map over the reflection on GMO (114) planes, displayed in Figure 2(b), shows a regular single peak positioned at $2\Theta = 73.69°$, which does not point to any twinning effects, in good agreement with the rather small thickness of α-layer.

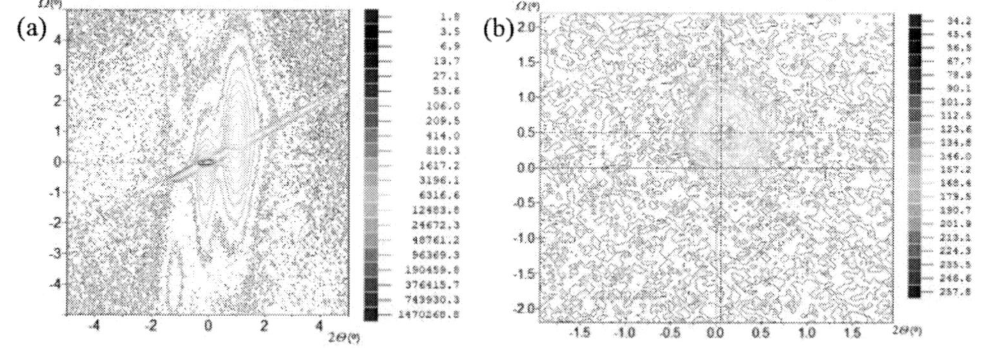

Fig. 2. (a) Symmetric Θ-Ω over substrate reflection (110) at $2\Theta = 32.42°$ and $\Omega = 16.21°$; (b) asymmetric (114) reflection at $2\Theta = 73.69°$, $\Omega = 19.78°$, $\Phi = 38.64°$ and $\Psi = -0.38°$ space maps.

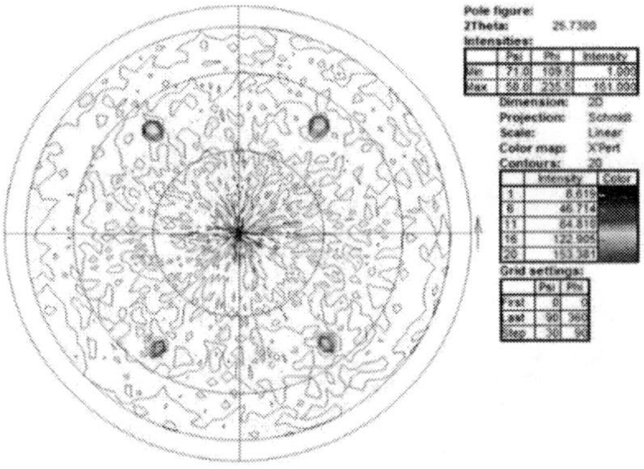

Fig. 3. Symmetric pole figure encompassing thin film (111) central reflection at $2\Theta = 25.73°$.

Table 1. Lattice parameters and relative variations of GMO thin film to STO substrate and bulk GMO.

STO substrate (Å)	$\Delta_{\gamma/STO}$ (%)	GMO γ-layer (Å)	$\Delta_{\gamma/GMO}$ (%)	GMO bulk (Å)
$\sqrt{2}.a_{STO}$ 5.530	< +1.4	a_γ 5.60(8)	< -4.4	a_o 5.866
$2.a_{STO}$ 7.820	> -2.3	b_γ 7.64(1)	> +2.8	b_o 7.431
		c_γ 5.37(2)	> +1.0	c_o 5.318
		V_γ 230.1(8)	< -0.7	V_o 231.81

Figure 3 shows the symmetric pole figure obtained at $2\Theta = 25.73°$ corresponding to reflection on the (111) planes of the GMO film. The observation of 4 reflections near $\Psi \sim 50.0° \pm 0.1°$ positioned at $\Phi = 55.2°$, 128.2°, 235.5° and 307.6°, confirms the orthorhombic symmetry of the GMO phase, and is compatible with the dominant off-plane preferential growth orientation of the film along c_γ-axis, from the γ-layer, with a_γ- and b_γ-axes being in-plane. The average projected angle $\Delta\Phi = 36.3° \pm 0.2°$ corresponds to the relation $a_\gamma/b_\gamma = 0.734(0)$. The calculated lattice parameters and the respective relative distortions to the STO substrate and bulk GMO are shown in Table 1, and appear as a compromise between both. Due to the reduced presence of α-layer, it was not possible to obtain the values of the in-plane b_α and c_α directly. Nonetheless, α-layer is highly constrained by epitaxial relations to either the STO(110) and the GMO γ-layer, corresponding to: $\sqrt{2}a_{STO} > c_\alpha \approx c_\gamma > c_O > a_\gamma$, and to $2a_{STO} > b_\alpha \approx b_\gamma > b_o$, which forcibly leads to an expansion of the lattice volume relative to the bulk GMO, estimated between +1.3% minimum and +6.8% maximum. Since γ-layer volume has a less pronounced variation of -0.7% from the bulk GMO, it suggests that α-layer grows on the STO substrate and acts as buffer for γ-layer.

Figures 4(a) and (b) show the real and imaginary part of the complex dielectric permittivity as a function of temperature, in the 20 kHz to 1 MHz frequency range. As temperature decreases, for frequencies above 100 kHz, the ε'(T) curves increase and reach a maximum at a frequency dependent temperature, while for frequencies below to 100 kHz this maximum is not reached down to 10 K. The ε''(T) curves monotonously increase as temperature decreases, with a strongly frequency dependent slope. These frequency dependent features are a clear evidence for a relaxation process that hinders the direct observation of anomalies associated with phase transitions. Figure 4(c) shows the Cole-Cole plots at different temperatures, which follow well defined semicircles at low temperatures, becoming gradually shorter for higher temperatures. The relaxation frequency was taken at the maximum of each semicircles, to obtain the Arrhenius plot shown in Figure 4(d). The data cannot be described by a single linear regime for the whole range. Two changes of slope occur, one between 35 K and 45 K, and another at around 20 K, associated with changes of the relaxation process activation energy. Above 45 K, the relaxation process activation energy is 7.3 ± 0.4 meV, between 35-45 K and 20 K it reduces to 1.8 ± 0.2 meV, while below 20 K it further drops to 0.46 ± 0.03 meV. Similar changes of the activation energy were also observed in bulk GMO, though from values of 10 to 20 meV, having been correlated with magnetic phase transitions [12]. Thus, we expect to have two phase transitions, one between 35 and 45 K, and another around 20 K. These temperatures are close to the two critical temperatures found in bulk GMO.

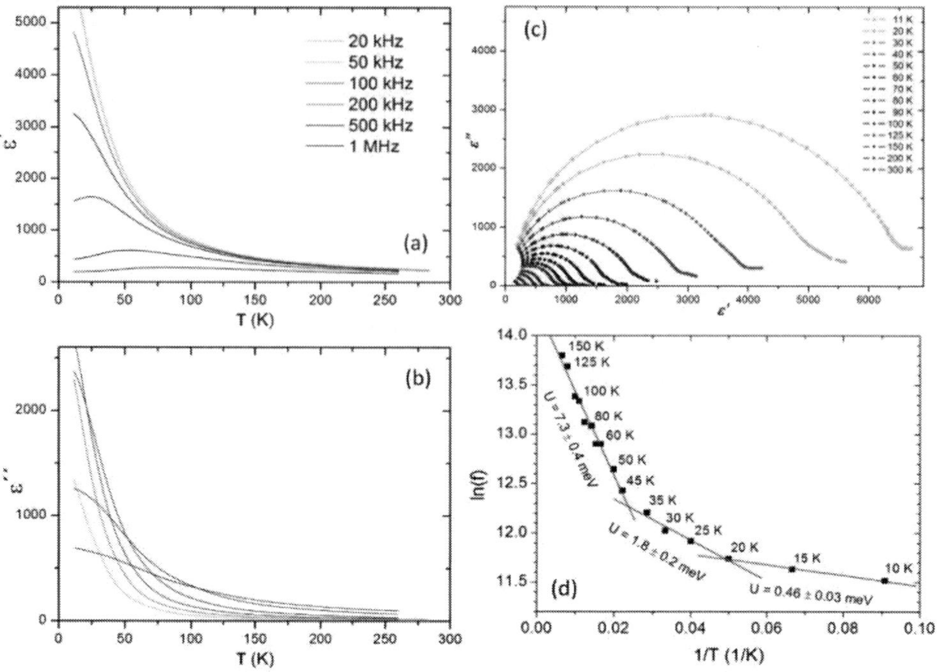

Fig. 4. (a) Real part and (b) imaginary part of complex dielectric permittivity measured in plane as a function of temperature for different fixed frequencies. (c) Cole-Cole plots at different fixed temperatures. (d) Arrhenius plot for the relaxation process with respective activation energies.

Figure 5 shows the magnetic response of the GMO film onto STO(110) substrate, measured in ZFC and FC conditions, under a 40 Oe magnetic field. The M(T) curve measured in ZFC conditions exhibits a peak-like anomaly at 20 K, at the same temperature where the change in the activation energy of the relaxation process is observed. The value of the M(T) increases significantly in FC conditions just below 25K. This result evidences for the stabilization of a canted AFM phase below T_C =20 K, which is responsible for the large increase of the separation between the ZFC and FC curves below this temperature. A faint but clear anomaly in M(T) measured in both ZFC and FC conditions is observed at 41 K, within the temperature interval of the second change in the activation energy. Although our data does not allow to unravel the nature of this anomaly, the stabilization of another magnetic phase cannot be excluded.

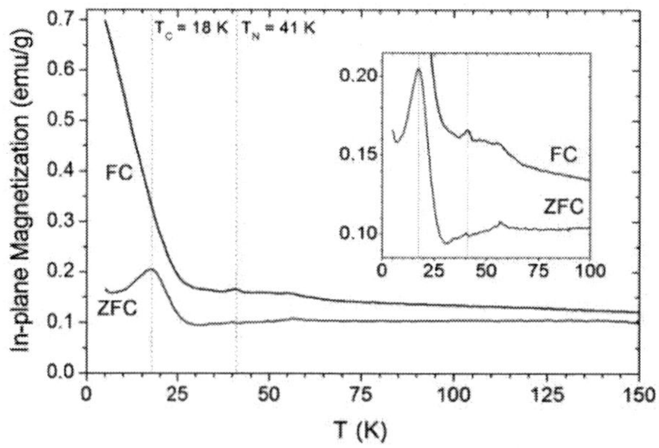

Fig. 5. In-plane magnetic response as a function of temperature, measured in ZFC and FC conditions with an applied field of 40 Oe. Vertical dashes lines mark phase transitions.

Although anomalies in the M(T) curves can also be seen at 55 K, we do not assign any magnetic phase transitions for two reasons: i) the relaxation process activation energy does not change at this temperature; ii) it has been reported that spurious oxygen found inside the SQUID magnetometer originates such type of anomalies around 55 K [13].

4 Conclusions

High quality GMO thin films were deposited onto (110)-oriented $SrTiO_3$ substrates by RF magnetron sputtering. The films exhibit an epitaxial orthorhombic symmetry, where the basal lattice parameters are strained by the cubic structure of the substrate, imposing a twinned α-phase on the GMO thin films in (100) direction that evolves to a γ-phase that grows preferably in (001) orientation. Both phases of the orthorhombic GMO thin film grown on STO(110) exhibit lattice parameters considerably less distorted in relation to the GMO bulk form than the tetragonal GMO thin film grown on STO(001). On the other hand, the volume compression of both the tetragonal and the main orthorhombic phases are noteworthy similar to -0.7% which denotes that this compression value is energetically favourable to the GMO perovskite phase under epitaxial strain. The temperature dependence of the relaxation process activation energy suggests for the existence of two magnetic phase transitions, also marked by anomalies in the M(T) curves. The magnetization data give clear evidence a canted AFM phase below Tc = 20 K.

References

1 W. Eerenstein, N.D. Mathur, J.F. Scott, Nature **442**, 759 (2006)

2 M. Bibes and A. Barthélémy, Nat. Mater. **7**, 426 (2008)

3 T. Kimura, T. Goto, H. Shintani, K. Ishizaka, T. Arima, Y. Tokura, Nature **426**, 55 (2003)

4 T. Kimura, S. Ishihara, H. Shintani, T. Arima, K.T. Takahashi, K. Ishizaka, and Y. Tokura, Phys. Rev. B **68**, 060403(R) (2003)

5 M. Mochizuki and N. Furukawa, Phys. Rev. B **80**, 134416 (2009)

6 T. Kimura, G. Lawes, T. Goto, Y. Tokura, A. Ramirez, Phys. Rev. B **71**, 224425 (2005)

7 T.Goto, Y.Yamasaki, H.Watanabe, T.Kimura, Y.Tokura, Phys. Rev. B **72**, 220403 (2005)

8 R. Vilarinho, A. Almeida, J.M. Machado da Silva, J.B. Oliveira, M.A. Sá, P.B. Tavares, J. Agostinho Moreira, Solid State Commun. **208**, 34 (2015)

9 X. Li, C. Lu, J. Dai, S. Dong, Y. Chen, N. Hu, G. Wu, M. Liu, Z. Yan, J.-M. Liu, Sci. Rep. **4**, 7019 (2014)

10 S. Mukherjee, K. Shimamoto, Y.W. Windsor, M. Ramakrishnan, S. Parchenko, U. Staub, L. Chapon, B. Ouladdiaf, M. Medarde, T. Shang, E.A. Müller, M. Kenzelmann, T. Lippert, C.W. Schneider, and. Niedermayer, Phys. Rev. B **98**, 174416 (2018)

11 P. Machado, F.G. Figueiras, R. Vilarinho, J.R.A. Fernandes, P.B. Tavares, M.R. Soares, S. Cardoso, J.P.B. Silva, A. Almeida, J.A. Moreira, Sci. Rep. accepted (2019)

12 R. Vilarinho, Magnetically-Induced Ferroelectricity Unraveled Through Spin-Phonon Coupling, University of Porto (2013)

13 D.G. Friend, NIST Stand. Ref. Databases volume 12 (1992)

Deposition of Lu-Fe-O thin films on silica glass substrates by MOCVD

A.F. Cardoso[1], *A*.A. Bassou[1], *V*.S. Amaral[2], *J*.R. Fernandes[3], and *P*.B. Tavares[1,*]

[1] CQVR, Chemistry Dep., ECVA, Univ. Trás-os-Montes e Alto Douro, 5001-801 Vila Real, Portugal

[2] CICECO, Physics Dep, Univ. Aveiro, 3810-193 Aveiro, Portugal

[3] CQVR, Physics Dep, ECT, Univ. Trás-os-Montes e Alto Douro, 5001-801 Vila Real, Portugal

Abstract. Thin films of the Lu-Fe-O system were deposited by aerosol assisted MOCVD on silica glass substrates. Hexagonal *h*-$LuFeO_3$, garnet $Lu_3Fe_5O_{12}$, perovskite o-$LuFeO_3$ or hematite Fe_2O_3 phases were obtained, depending on the thermodynamic deposition conditions or post annealing temperature. Magnetic measurements confirm the ferromagnetic behaviour at room temperature of the thin films with garnet phase. An indirect bandgap of 1.78 eV was measured.

1 Introduction

Recently, the Lu-Fe-O system has received a great deal of attention due to the multiferroic properties of several phases of the system [1, 2]. The garnet phase, $Lu_3Fe_5O_{12}$, besides being ferromagnetic at room temperature, presented magnetodieletric effect[3]. The hexagonal $LuFeO_3$ phase shows multiferroic properties at room-temperature [4].

The ½Lu_2O_3-½Fe_2O_3-FeO phase diagram is presented in figure 1a), following the sketches by Sekine [5], Kumar [6] and Wang [7]. According to this diagram it is important to consider the mixed valence states Fe(II)/Fe (III), adjustable by oxygen partial pressures during sample preparation. With higher oxygen partial pressures there are the Fe_2O_3 (hematite) and Lu_2O_3 terminal phases, plus the intermediate $Lu_3Fe_5O_{12}$ (garnet), *o*-$LuFeO_3$ (orthorhombic perovskite) and *h*-$LuFeO_3$ (hexagonal). The $LuFeO_3$ structure exhibits distortion from ideal perovskite due to the small ionic rare earth radius. This compound has two different structures: a stable P*nma* orthorhombic perovskite (*o*-LuFeO3) and a metastable hexagonal phase P*6₃cm* (*h*-LuFeO3). Decreasing the oxygen partial pressure, the phases Fe_3O_4 (magnetite), $LuFe_2O_4$ and $Lu_2Fe_3O_7$ appear. The other terminal phase of this diagram is FeO. The $LuFe_2O_4$ and $Lu_2Fe_3O_7$ phases are also metastable [7]. Peritetic decompositions of these phases give rise to *o*-$LuFeO_3$ and $Lu_3Fe_5O_{12}$ [8].

Wang suggests a Van't Hoff diagram for this system, as a result of his studies during PLD deposition of thin films on MgO (111) substrates [7]. According to this diagram it would be possible to obtain a mixture between the *h*-$LuFeO_3$ and $LuFe_2O_4$ phases by controlling the oxygen partial pressure around 2.7×10^{-5} mbar at 900°C.

* Corresponding author: ptavares@utad.pt

In the present study, we have deposited thin films of the Lu-Fe-O system, with different compositions and at different temperatures, by MOCVD, on silica glass substrates. These substrates are much less expensive than the traditional monocrystalline ones, like sapphire, YSZ, MgO, $SrTiO_3$ or $LaAlO_3$, among others. However, no epitaxy effects, enhancing the growing of a specific phase, are expected. The films were post annealed in air at increasing temperatures to investigate the thermodynamic stability of the phases. Structural characterizations were performed by X-ray diffraction. Further characterizations were performed by magnetic and optical measurements.

2 Experimental procedures

In the deposition of thin films by MOCVD[9], metalorganic precursors, $Lu(tmhd)_3$ and Fe(tmhd)3, were synthesized from tmhd (2,2,6,6-tetramethyl-3,5-heptanedione) using the method described by Eisentraut and Severs [10] and purified by sublimation. These precursors were dissolved in bis-2-methoxyethylether (concentration from 2.5 up to 8.5 μM) and ultrasonic nebulized at 810 kHz to produce an aerosol with 3 μm mean-particle diameter. Argon was used as a carrier gas (40 mL/min) and heated at 260 °C to volatilize both solvent and metalorganics. Preheated oxygen (160 mL/min) was also mixed. The substrates were glued to the stainless-steel susceptor using silver paste and induction heated between 500°C and 850°C. Total pressure was 9 mbar (1.8 mbar of argon and 7.2 mbar of oxygen). The deposition time was 1 hour to produce a film thickness around 250 nm.

X-ray diffraction patterns were acquired on a Panalytical X'Pert Pro MPD equipped with a X'Celerator detector and secondary monochromator, in Bragg-Bentano geometry. A CuKα wavelength, with λ=1.5418Å, a step size of 0.017° with 100s/step was used. Rietveld refinement of the diffractograms was performed using PowderCell software [11], allowing phase quantification.

Scanning Electron Microscopy (SEM) and Energy Dispersive Spectrometry (EDS) were performed using a FEI Quanta 400 with W filament and EDAX system respectively.

Thermal treatments of the films were conducted on a tubular oven with a calibrated K thermocouple located just above the thin film.

The UV/Vis transmission spectra of the thin films were acquired using LLG UNISpec 2, from 200 up to 1100 nm. Direct and indirect bandgaps were calculated through Tauc plots [12, 13].

The UV/Vis spectra were simulated using the Swanepoel model [14] to obtain the film thickness, refractive index and absorption coefficient. According to this model, the experimental $T(\lambda)$ transmittance spectrum obtained for a thin film depends on the optical properties of the film and substrate.

Magnetic measurements were conducted using a SQUID Magnetometer from Quantum Design MPMS3, with a superconductor coil of up to 7 Tesla and temperatures from 1.8K up to 400K.

3 Results and discussion

In this work we selected four compositions of the Fe(III) part of phase diagram of fig. 1a), to perform depositions by MOCVD. These compositions are marked as A, B, C and D in fig. 1b). Composition A corresponds to Fe_2O_3, composition B to a region between Fe_2O_3 and $Lu_3Fe_5O_{12}$, region C to $Lu_3Fe_5O_{12}$ and composition D to a region between $Lu_3Fe_5O_{12}$ and $LuFeO_3$.

 https://doi.org/10.1051/epjconf/202023305006

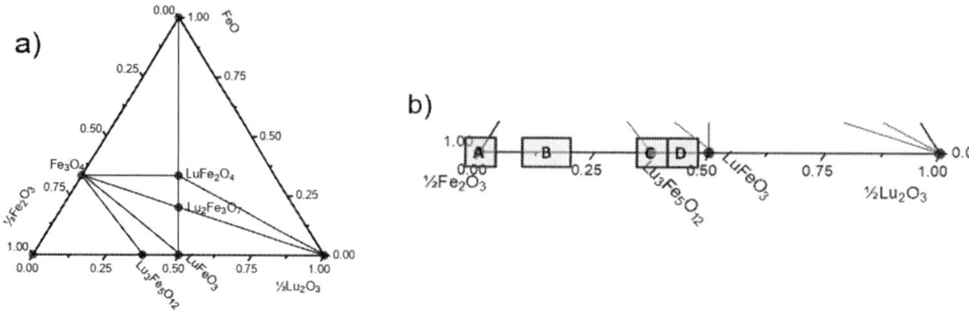

Fig. 1 – a) Phase diagram of the ½ Fe_2O_3-½ Lu_2O_3-FeO system [5-7]; b) Fe(III) region of the Lu-Fe-O phase diagram representing the composition of the deposited films.

3.1 Films with composition A

The composition A corresponds to films deposited by MOCVD using only Fe(tmhd)3 as metalorganic precursor. In fig. 2 a) we present the X-ray diffraction patterns of the films as deposited at temperatures from 500 up to 750°C together with expected diffraction pattern of Fe_2O_3 hematite [15]. The wide peak around 22° is due to the silica glass substrate. The (104) peak of hematite, the most intense one, appears as early as 500°C indicating that phase formation starts at this temperature. The (110) peak, the second most intense, appears at 650°C indicating a well crystallized phase. As expected from a deposition on an amorphous substrate, no preferable orientation is observed. At higher deposition temperatures no substantial changes in the diffraction patterns can be seen.

3.2 Films with composition B

The composition of these films is in the Lu/(Lu+Fe)=0.2-0.3 (atomic) range, below the garnet composition of Lu/(Lu+Fe)=0.375. In fig. 2b) we see that a thin film deposited at 700°C is amorphous. Ex-situ thermal treatment at 850°C during 1 hour, promotes the crystallization of the film yielding mainly hexagonal *h*-$LuFeO_3$ phase and traces of garnet and hematite phases. At 900°C the *h*-LuFeO3 phase starts to decompose and no traces of this phase is found at 950°C. In the opposite way, the garnet and hematite phases dominate at these temperatures, according to what is expected from the analysis of the phase diagram of fig. 1. However, some peaks of perovskite phase can also be detected, probably due to the peritetic decomposition of the hexagonal phase, indicating that the film is not in thermodynamic equilibrium. A small peak at 21.8° is due to the crystallization of the silica glass substrate into cristobalite phase [16].

3.3 Films with composition C

The composition of this film match the Lu/(Lu+Fe)=0.375 atomic ratio and was deposited by MOCVD at 800°C and consecutively thermal treated ex-situ at 825°C and 850°C during 1 hour. The corresponding diffraction patterns are displayed in fig. 2 c). At 800°C the film is mainly amorphous, with a trace of a broad (004) peak of *h*-$LuFeO_3$ phase at 30.9° and a narrow garnet $Lu_3Fe_5O_{12}$ phase (420) peak at 32.7°. The thermal treatment at 825°C promotes the crystallization of the garnet phase, which is well crystallized at 850°C.

3.4 Films with composition D

A film with Lu/(Lu+Fe)≈0.4 atomic ratio was deposited at 700°C and consecutively thermal treated with 50°C steps, during 1 hour, up to 1000°C (fig. 2d). The as-deposited film was mainly amorphous with an incipient (004) peak of the hexagonal phase at 30.8°. The crystallization of this phase occurs at 800°C with the appearance of the (002) peak at 15.2°. At 900°C the hexagonal phases decomposes and the garnet and perovskite phases crystallized. At 1000°C a new phase appears, monoclinic $Lu_4Si_4O_{14}$ [17], resulting from the interdiffusion of the lutetium to the substrate.

The X-ray diffraction patterns were deeply analysed by Rietveld refinement, allowing phase quantification. The silica glass structure was simulated by Le Bail method [18]. The results are displayed in fig. 3 a). To go deeper into the phase formation, the results of the Lu-Fe-O phases (hexagonal, garnet and perovskite) were normalized to 100% and plotted in fig. 3b). In this figure, we observed the decomposition of the hexagonal phase between 900 and 950°C and at the same time the increase of the garnet and perovskite phases.

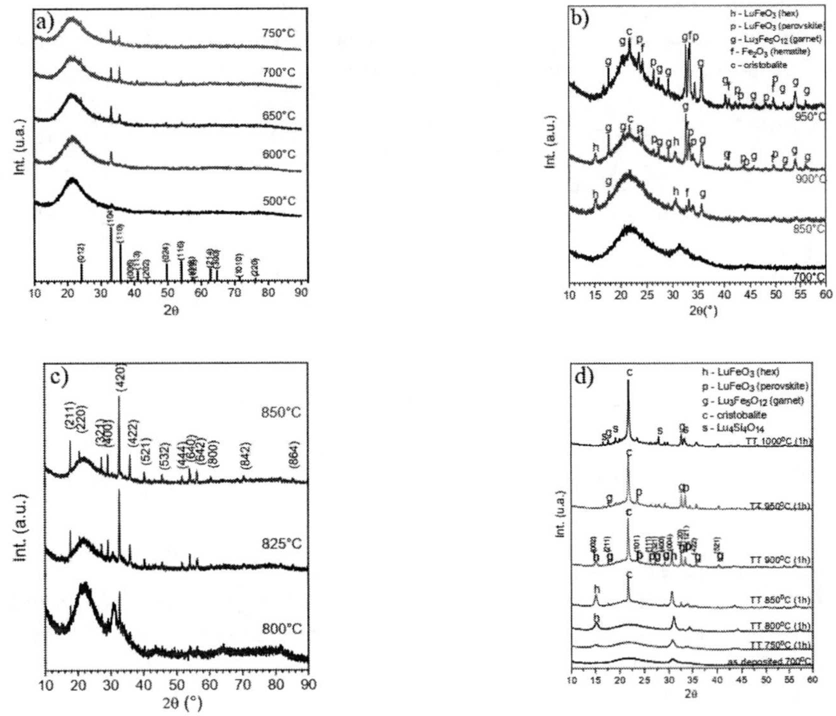

Figure 2 – Temperature effects on the X-ray diffraction patterns of thin films deposited by MOCVD with different compositions: a) Composition A; b) Composition B; c) Composition c); Composition d).

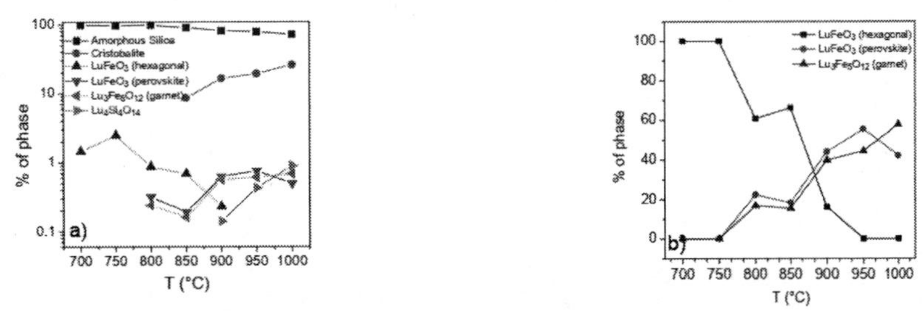

Figure 3 – Phase quantification results from Rietveld refinement of film with D composition; a) results with all identified phases including silica glass amorphous substrate; b) normalized results with only the Lu-Fe-O phases.

3.5 Magnetic measurements

For the film with composition B, thermal treated at 900°C, presenting the garnet phase, we have measured the temperature dependence of magnetization M(T) at 100Öe magnetic field, with increasing temperature up to 400 K, after a zero field cool down to 3.6 K (fig.

4a). For this measurement we have considered a film thickness of 283 nm, determined by the Swanepoel model (fig. 5a), a substrate dimension of 5.9x5.9 mm^2, therefore a volume of $9.85x10^{-6}$ cm^3. An average density of 6.122 gcm^{-3} was estimated, resulting in a $6.0x10^{-5}$ g film mass. The film presents a magnetization of 3.4 emu/g at 4K and 2.5 emu/g at room temperature. The Curie temperature is clearly well above 400K. Previous published results of the $Lu_3Fe_5O_{12}$ phase indicate that Curie temperature is around 550 K [19]

M(H) hysteresis loop, made at room temperature, is presented in fig. 4b). A typical soft ferromagnetic loop is observed, with a small contribution from a diamagnetic phase due to the silica glass substrate. No exchange bias is detected. The coercive field is H_c=85Öe. A saturation magnetization of M_s=4.2 emu/g was measured, which is 30% of the reported value of 14 emu/g for a pure phase ceramic sample [20]. The amount of the garnet phase calculated by Rietveld refinement from the X-ray diffraction pattern is 29 wt%. We can consider that the hexagonal and the perovskite phases do not contribute to the magnetic moment.

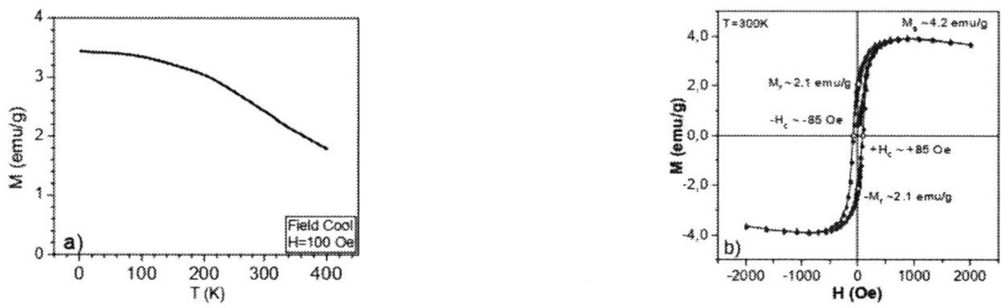

Figure 4 – Magnetic measurements of film with B composition thermal treated at 900°C. a) M(T) at 100Öe; b) M(H) at 300K.

3.6 Optical measurements

The UV/Vis spectra of the films were simulated by Swanepoel model. As a representative case, in fig. 5a) we present the results obtained with film of B composition. The estimated film thickness is 283±1 nm.

To our knowledge, the band structure is not published for these materials, so we do not know if the transition is direct or indirect. Direct and indirect bandgaps are calculated from Tauc plots in fig. 5b) and 5c) respectively. Heat treatment at 950°C decreases the direct band gap from 2.43 eV (for films deposited at 700°C) to 2.13 eV. These values, for a direct transition, agree with those published for hematite films, around 2.2 eV [21].

In the case of indirect transitions, there are two energies, as indicated by Misho, the highest (1.78 eV) being attributed to the band gap and the lowest (0.95 eV) to the energy required for the phonon. [22].

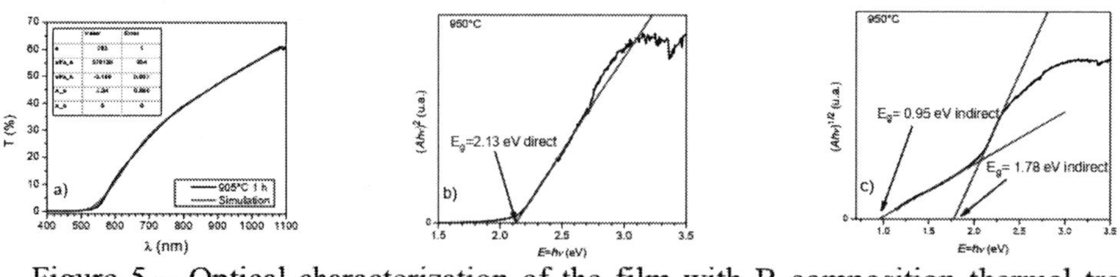

Figure 5 – Optical characterization of the film with B composition thermal treated at 950°C. a) UV/Vis spectra and simulation with Swanepoel model; b) Tauc plot for direct bandgap determination; c) Tauc plot for indirect bandgap determination.

The authors would like to acknowledge Fundação para a Ciência e Tecnologia (FCT) through projects CERN/FIS-PAR/0005/2017, NORTE/01/0145/FEDER/028538,

FEDER/PT2020, projects UID/QUI/00616/2013 and CICECO-Aveiro Institute of Materials, FCT Ref. UID/CTM/50011/2019. A.A. Bassou acknowledges scholarship of the project NORTE/01/0145/FEDER/028538.

4 Conclusions

Thin films of the Lu-Fe-O were successfully deposited by aerosol assisted MOCVD from β-diketones Lu(tmhd)$_3$ and Fe(tmhd)$_3$ on silica glass substrates. Depending on the composition and on the deposition or thermal treatment temperatures, hematite, $Lu_3Fe_5O_{12}$ garnet, hexagonal $LuFeO_3$ or orthorrombic $LuFeO_3$ perovskite phases are detected on the films by X-ray diffraction. A film with garnet phase presented ferromagnetic behaviour with M_s= of 4.2 emu/g and H_c=85Öe at room temperature. Optical measurements indicate a direct bandgap transition with 2.13 eV or an indirect bandgap of 1.78 eV.

References

1 - A.R. Akbashev, V.V. Roddatis, A.L. Vasiliev, S. Lopatin, V.A. Amelichev, A.R. Kaul, Scientific Reports **2**, 672 (2012).
2 – J.A. Mundy, C.M. Brooks, M.E. Holtz, J.A. Moyer, H. Das, A.F. Rébola, J.T. Heron, J.D. Clarkson, S.M. Disseler, Z. Liu, A. Farhan, R. Held, R. Hovden, E. Padgett, Q. Mao, H. Paik, R. Misra, L.F. Kourkoutis, E. Arenholz, A. Scholl, J.A. Borchers, W. D. Ratcliff, R. Ramesh, C.J. Fennie, P. Schiffer, D.A. Muller, D.G. Schlom, Nature **523**, 537 (2016).
3 - Ying Hou, J. Phys. D: Appl. Phys. **51**, 275001 (2018).
4 - Wenbin Wang,, Jun Zhao, Wenbo Wang, Zheng Gai, Nina Balke, Miaofang Chi, Ho Nyung Lee, Wei Tian, Leyi Zhu, Xuemei Cheng, David J. Keavney, Jieyu Yi, Thomas Z. Ward, Paul C. Snijders, Hans M. Christen, Weida Wu, Jian Shen, Xiaoshan Xu, Physical Review Letters **110**, 237601 (2013).
5 – T. Sekine, T. Katsura, J. Solid State Chem., **17**, 49-54 (1976).
6 - M.S. Vijaya Kumar, Kazuhiko Kuribayashi, Koichi Kitazono, J. Mater. Res., **23**. 11, (2008).
7 - Wenbin Wang,, Zheng Gai,, Miaofang Chi, Jason D. Fowlkes, Jieyu Yi, Leyi Zhu, Xuemei Cheng, David J. Keavney, Paul C. Snijders, Thomas Z. Ward, Jian Shen, Xiaoshan Xu, Physical Review B **85**, 155411 (2012).
8 - Y. Qin, X.Q. Liu, Y. Wu, X.M. Chen, J. of the American Ceramic Society, **96**, 2506 (2013).
9 - P.B. Tavares, V.S. Amaral, J.P. Araújo, J.B. Sousa, J.M. Vieira, J. Appl. Phys., **85**, 8, (1999).
10 - K. Eisentraut, R. Sievers, Journal of the American Chemical Society, **87**, 22, 5254 (1965).
11 - W. Kraus, G. Nolze, PowderCell for Windows, version 2.4, available at http://www.ccp14.ac.uk/tutorial/powdcell/
12 - J. Tauc, Materials Research Bulletin. **3**, 37 (1968).
13 - M.N. Batin, V. Popescu, Optoelectronics and Advanced Materials, **6,** 7-8, 727 (2012).
14 - R. Swanepoel, Journal of Physics E: Scientific Instruments, **16**, 1214 (1983).
15 – American Mineralogist Crystal Structure Database, file number 0017806.
16 - American Mineralogist Crystal Structure Database, file number 0001629.
17 – Crystal Open Database, file number 8100597
18 - Le Bail, J. of Non-Crystalline Solids, **183**, 39 (1995).
19 - P. Manimuthu, R. Vidya, P. Ravindran, H. Fjellvåg, C. Venkateswaran, Physical Chemistry Chemical Physics, **17**, 27, 17688 (2015).
20 - X. Wu, X. Wang, Y. Liu, W. Cai, S. Peng, F. Huang, X. Lu, F. Yan, J. Zhu, Applied Physics Letters, **95**, 182903, (2009).
21 - B. Gilbert, C. Frandsen, E. Maxey, D.M. Sherman, Physical Review B, **79**, 3 (2009).
22 - R.H. Misho, W.A. Murad, Solar Energy Materials and Solar Cells, **27**, 335 (1992).

Low temperature characterization of high efficiency spin-filter Josephson junctions

Roberta Caruso[1,2,3,*], *Halima Giovanna* Ahmad[1,2], *Avradeep* Pal[4,5], *Giovanni Piero* Pepe[1,2], *Davide* Massarotti[6,2], *Mark G.* Blamire[5], and *Francesco* Tafuri[1,2]

[1]Dipartimento di Fisica E. Pancini, Università degli Studi di Napoli Federico II, Monte S. Angelo, via Cinthia, I-80126 Napoli, Italy
[2]CNR-SPIN, c/o Complesso di Monte S. Angelo, via Cinthia, I-80126 Napoli, Italy
[3]SeeQC-eu, via dei Due Macelli 66, I-00187 Roma, Italy
[4]Indian Institute of Technology Bombay, Main Gate Rd, IIT Area, Powai, Mumbai, Maharashtra 400076, India
[5]Department of Materials Science and Metallurgy, University of Cambridge, 27 Charles Babbage Road, Cambridge CB3 0FS, United Kingdom
[6]Dipartimento di Ingegneria Elettrica e delle Tecnologie dell'Informazione, Università degli Studi di Napoli Federico II, via Claudio, I-80125 Napoli, Italy

Abstract. The interplay between superconducting and ferromagnetic order parameters in S/F interfaces gives rise to a wide range of peculiar properties with applications in high-efficiency computation and in the emerging field of superconducting spintronics. In NbN/GdN/NbN Josephson junctions, GdN barriers give unique properties due to the double insulting and ferromagnetic nature of the material, as demonstrated in previous works. Here we focus on tunneling spectroscopy of these junctions down to 0.3 K when changing the barrier thickness, which contributes to complete a consistent picture on the physics of these junctions and supports the previous indications of equal-spin Cooper pairs contributing to the total supercurrent of the devices.

1 Introduction

In the last twenty years, the advancements in thin film deposition have made possible the realization of superconductor/ferromagnet heterostructures, and a subsequent rich literature on the experimental characterization and the theoretical modeling of such systems. However, most of this literature focuses on metallic junctions, employing a wide range of ferromagnets as weak links. Spin filter Josephson junctions are characterized by a barrier with the unique double nature of ferromagnet and insulator. These junctions have been predicted to have a number of interesting properties for applications in the field of low dissipation quantum and classical computation, such as π-shifts with intrinsically low dissipation [1, 2] and unconventional temperature dependence of the critical current I_C [3]. The first realization of spin-filter NbN/GdN/NbN Josephson junctions, together with an unconventional $I_C(T)$ down to 4.2 K has been reported in [4]. Evidences of an anomalous $I(\phi)$ compatible with triplet correlations have been shown in [5]. Further demonstrations of the anomalous properties of these systems

*e-mail: caruso@fisica.unina.it

have been reported recently [6, 7]. Here we present conductance measurements confirming the previous indications of equal-spin coupling inside the barrier.

2 Methods

The junctions used in this work are NbN/GdN/NbN junctions with different barrier thicknesses, ranging from 1.5 nm to 4 nm, realized during the same deposition run to ensure the same deposition conditions for all samples, using methods described elsewhere [5]. Measurements have been performed using an Heliox VL evaporation cryostat with base temperature of 0.3 K, with customized low noise filter detailed in [8, 9].

One of the peculiarities of the ferromagnetic junctions with insulating barriers is the spin filtering property. The presence of an exchange field due to the ferromagnet gives rise to two different barrier heights for spin-up and spin-down electrons. In this case, electrons seeing a higher barrier have an exponentially lower tunneling probability, and so the current passing through the junction is spin-polarized. The spin-polarization P can be calculated from experimental $R(T)$ curves using the approximate relation $P \approx \tanh\left[\cosh^{-1}\left(\frac{\sigma}{\sigma^*}\right)\right]$, where σ is the measured conductance and σ^* is the conductance in absence of exchange field [4]. In Fig. 1, the black dots are the measured resistance $R \propto 1/\sigma$, while the red lines are obtained by fitting experimental $R(T)$ above T_{Curie} and then extrapolating R^* below T_{Curie}.

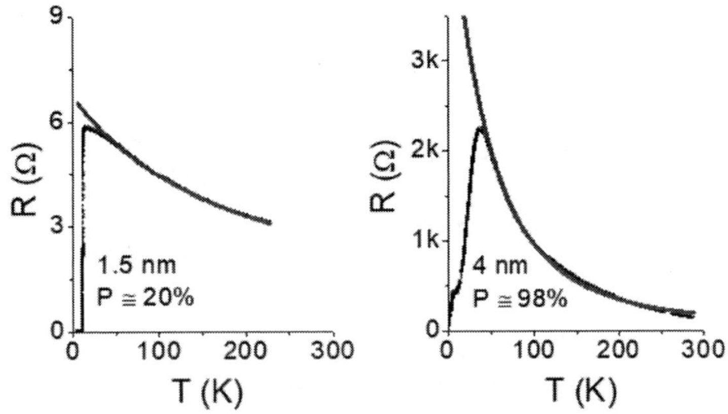

Figure 1. $R(T)$ curves for samples with different barrier thicknesses. Black dots are experimental points, red line is the resistance obtained by fitting data with a standard semiconducting model below T_{Curie}. Left: junction with low spin polarization. Right: junction with high spin polarization, the resistance drop below T_{Curie} is clearly visible.

In order to measure the conductance spectra the junctions are current biased with a sum of two signals: a low frequency (1mHz) triangular ramp, and a small sinusoidal excitation at about 30 Hz. The optimal ratio between the two signal amplitudes is 10^{-3}. The output dV is read using a lock-in amplifier, while dI corresponds to the amplitude of the sinusoidal excitation. The quasi-dc output due to the low frequency component of the bias is read using a standard multimeter. Details of the procedure can be found elsewhere [10]. A small magnetic field is applied to suppress the Josephson current in the junction, such field is chosen so that it correspond to one of the minima of the $I_C(H)$ pattern, and it is much smaller than the critical field of GdN, so it does not affect significantly the main features of the conductance spectra attributed to spin filtering.

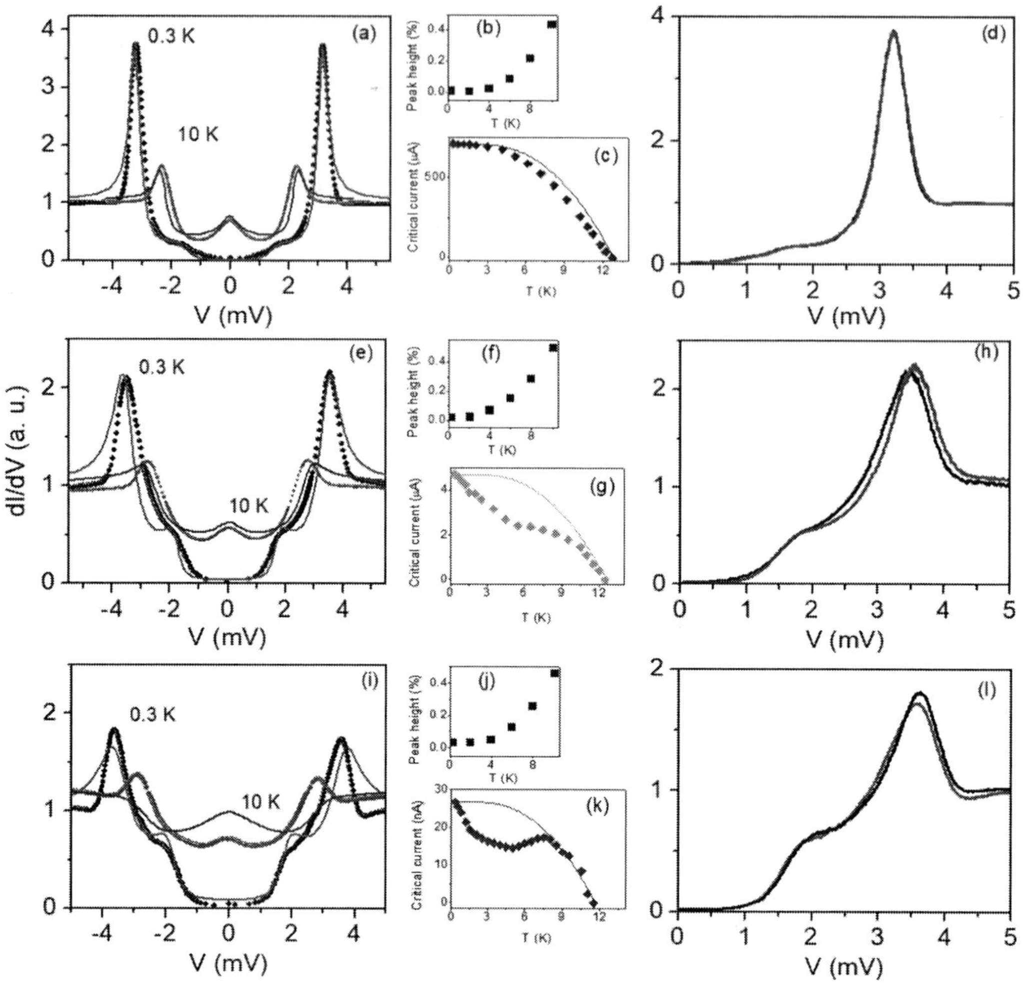

Figure 2. Conductance spectra for junctions with $P \leq 20\%$ (a), $P \approx 97\%$ (e) and $P \approx 99\%$ (i) at 0.3 K and at 10 K. The temperature evolution of the peak at zero bias is shown in (b), (f) and (j) respectively. The peak height is reported on Y axis as the ratio between the zero bias peak height and the height of the peak corresponding to the superconducting gap. Panels (d), (h) and (l) show the overlap of the positive and negative branch of the curves at 0.3 K for the three samples mentioned above. The shift between the two branches is proportional to the ferromagnetic exchange energy. In panels (c), (g) and (k) we show the critical current as a function of temperature (diamonds), in comparison with the standard behavior of Josephson junctions with insulating barriers.

3 Conductance measurements

Measurements of IV curves as a function of temperature highlighted an anomalous $I_C(T)$ behavior for junctions with barrier thicknesses above 3 nm, which has been extensively discussed in [7]. In that work, the relevance of the magnetic activity of the barrier is discussed, and a phenomenological model to describe the anomalous $I_C(T)$ curves is detailed. Here we present measured dI/dV curves and their relation with the presence of triplet coupling inside the barrier.

In Fig. 2 (a), (e) and (f) are shown the conductance spectra for three junctions with different spin filter efficiencies (<20%, 97% and 99%) for two different temperatures. For all the samples we observe the appearance of two symmetric structures at $V \approx \Delta$ at low temperatures. Similar structures are observed in the presence of a Zeeman splitting that shifts the spin bands of the superconductor [11]. Since the subgap structures become increasingly pronounced as the spin polarization increases, it is reasonable to assume for our samples that these are due to the exchange field induced in the superconducting electrodes by the GdN barrier. Furthermore, analogous experiments in NbN/GdN/TiN (where TiN is a non-superconducting metal) do not show any sign of subgap structures, while they show evidence of an exchange field induced in the NbN superconducting electrode [12]. This implies that the structures observed in NbN/GdN/NbN Josephson junctions should be related to the exchange fields induced by GdN in each of the superconducting electrodes.

In addition, we observe a zero-bias peak starting at 4 K up to 10 K [see Fig. 2 (b), (f) and (j)], and a finite conductance at zero voltage at all temperatures, in contrast with the typical behavior of a standard BCS junction. Similar features have been reported in NbN/GdN/TiN junctions, where the zero-bias peak has been related to the presence of odd-frequency superconducting order [13]. However, since the TiN layer is metallic, the Josephson coupling does not occur and consequently no triplet supercurrent can be observed. In these junctions the zero-bias peak corresponds to the situation where the superconducting density of states (DOS) is enhanced at the Fermi level. Similarly, in NbN/GdN/NbN Josephson junctions, the zero-bias peak should correspond to an unconventional behavior of the DOS of the superconducting electrodes in the vicinity of the Fermi level. In order to account for the presence of available states at the Fermi level, one can use the following modified expression for the density of states:

$$\rho(E) = \frac{1}{c+1}\left(Re\left[\frac{|E - i\Gamma|}{\sqrt{(E - i\Gamma)^2 - \Delta}}\right] + c\right) \tag{1}$$

The fitting parameter Γ accounts for the finite lifetime of the Cooper pairs [14] and it has been used in plenty of experiments to explain the deviation of the measured curves from the standard exponential subgap of BCS superconductors [15]. As the spin-filter efficiency increases, we find that the subgap region dI/dV curves becomes almost polynomial instead, a feature that is reproduced in the simulations by increasing the smearing Γ, thus making it possible to correlate the shape of the dI/dV curve with the increasing magnetic activity of the samples. In analogy with the results on NbN/GdN/TiN junctions we include the second fitting term c, since Γ alone is not sufficient to account for the finite background conductance observed experimentally [12, 13]. The conductance formula can be then obtained deriving the following expression for the quasiparticle Josephson current [16–18]:

$$\begin{aligned} I \propto a \int_{-\infty}^{+\infty} \rho_1(E+h)\rho_2(E-h+V)\,[f(E) - f(E+V)] \,+ \\ b \int_{-\infty}^{+\infty} \rho_1(E-h)\rho_2(E+h+V)\,[f(E) - f(E+V)] \end{aligned} \tag{2}$$

where ρ_1 and ρ_2 are the DOS in the two superconducting electrodes, $f(E)$ is the Fermi function, a and b are coefficient used to model spin filter efficiency and h is the exchange field of the ferromagnet. The coefficient a is the spin filter efficiency, estimated from $R(T)$ curves, while b is chosen so that $a + b = 1$.

As we stated before, the exchange field h is known to cause the Zeeman splitting of the spin bands in the superconductor, and the appearance of characteristic subgap features in dI/dV curves of superconductor/insulator/normal metal junctions. However, in the case of very high spin polarization one of the two bands is almost completely suppressed, and thus the main signature of the Zeeman splitting in dI/dV curves is a voltage shift corresponding to the actual exchange energy induced in the superconductor, as observed in NbN/GdN/TiN junctions [12, 13]. In order to estimate h, we analyzed the symmetry of the conductance spectra of samples with different barrier thicknesses. In Fig. 2 (d), (h) and (l), the positive and negative branch of the differential conductance are superimposed. For the junction with lowest spin filter efficiency the two branches overlap, indicating the absence of a significant exchange field. Assuming an exchange energy with the same magnitude but opposite sign in the two electrodes, the difference between the positive and negative peaks of the conductance is proportional to h.

The magnetic field corresponding to the exchange energy h is given by:

$$\mu_0 H = \frac{2eh}{g\mu_B} \tag{3}$$

where μ_0 is the vacuum permittivity, e is the electron charge, g is the Lande factor, which is 2 for GdN, and μ_B is the Bohr magneton. For the junctions shown in Fig. 2 (h) and (l) the values obtained for the exchange field are respectively $\mu_0 H = 0.43 \pm 0.2T$ and $\mu_0 H = -0.17 \pm 0.1T$. These values are far larger than the typical values of the Fraunhofer modulation in these junctions. The sign of the exchange field is inverted between the two junctions with high spin polarization and the absolute value decreases with increasing barrier thickness. This last result is in agreement with previous observations of a decreasing barrier height with increasing GdN thickness, due to the presence of two Schottky barriers at the interfaces, rather than a conventional square barrier, while the sign inversion of the exchange field H can be attributed to the changes in the unsaturated GdN domain structure occurring at larger thicknesses [19].

The fitting curves in Fig. 2 show that Eq. 2 reproduces the experimental data with good approximation, thus supporting the hypothesis of an unconventional DOS at the Fermi level related to the exchange fields induced in the superconducting electrodes. In junctions with high spin polarization, Eq. 2 describing the quasi-particle current gives a very rough reproduction of the measured dI/dV curve. This result can be compared with the one reported in [7], where the $I_C(T)$ behavior can be only roughly reproduced without considering a much more complex magnetic structure of the barrier, including a strong asymmetry in the induced exchange fields. These experimental indications of an increasing asymmetry in the exchange fields induced in the NbN electrodes coming from both dI/dV measurements and $I_C(T)$ curves [Fig. 2 (c), (g) and (k)] are compatible with the presence of spin-triplet transport mechanisms contributing to the total supercurrent in spin-filter Josephson junctions.

4 Conclusion

The previous observations of anomalous $I_C(T)$ curves for high spin filter efficiencies imply the presence of spin-triplet supercurrents in NbN/GdN/NbN Josephson junctions. The subgap structures and the zero bias peak observed in the conductance spectra analyzed in this work are attributed to an asymmetric induced exchange field in the superconducting electrodes and to a finite background conductance caused by a non-conventional density of states in the superconducting electrodes compatible with equal-spin Cooper pairs contributing to the total current.

References

[1] S. Kawabata, S. Kashiwaya, Y. Asano, T. Tanaka, and A. A. Golubov, Phys. Rev. B **74**, 180502 (2006).

[2] S. Kawabata, Y. Asano, T. Tanaka, A. A. Golubov, and S. Kashiwaya, Phys. Rev. Lett. **104**, 117002 (2010).

[3] M. Fogelstrom, Phys. Rev. B **62**, 11812-11819 (2000).

[4] K. Senapati, M. G. Blamire, Z. H. Barber, Nature Materials **10**, 849-852 (2012).

[5] A. Pal, Z. H. Barber, J. W. A. Robinson and M. G. Blamire, Nature Communications **5**, 3340 (2014).

[6] D. Massarotti, A. Pal, G. Rotoli, L. Longobardi, M. G. Blamire and F. Tafuri, Nature Communications **6**, 7376 (2015).

[7] R. Caruso, D. Massarotti, G. Campagnano, A. Pal, H. G. Ahmad, P. Lucignano, M. Eschrig, M. G. Blamire and F. Tafuri, Phys. Rev. Lett. **122**, 047002 (2019).

[8] L. Longobardi, D. Massarotti, G. Rotoli, D. Stornaiuolo, G. Papari, A. Kawakami, G. P. Pepe, A. Barone and F. Tafuri, Appl. Phys. Lett. **99**, 062510 (2011).

[9] D. Stornaiuolo, G. Rotoli, D. Massarotti, F. Carillo, L. Longobardi, F. Beltram, and F. Tafuri, Phys. Rev. B **87**, 134517 (2013).

[10] D. Stornaiuolo, D. Massarotti, R. Di Capua, P. Lucignano, G. P. Pepe, M. Salluzzo, and F. Tafuri, Phys. Rev. B **95**, 140502 (2017).

[11] R. Meservey, P. Tedrow, and P. Fulde, Phys. Rev. Lett. **25**, 1270 (1970).

[12] A. Pal and M. G. Blamire, Phys. Rev. B **92**, 180510(R) (2015)

[13] A. Pal, J. A. Ouassou, M.Eschrig, J. Linder and M. G. Blamire, Scientific Reports **7**, 40604 (2016).

[14] R. C. Dynes, V. Narayanamurti, and J. P. Garno, Phys. Rev. Lett. **41**, 1509 (1978).

[15] E. L. Wolf, *Principles of Electron tunneling spectroscopy* (Oxford University Press, Oxford, 2nd edition, 2011)

[16] R. Meservey and P. M. Tedrow, Phys. Rev. B **11**, 4224-4236 (1975).

[17] X. Hao, J. S. Moodera, R. Meservey, Phys. Rev. B **42** 8235-8244 (1990).

[18] P. M. Tedrow, J. E. Tkaczyk and A. Kumar, Phys. Rev. Lett. **56**, 1746-1749 (1986).

[19] A. Pal, K. Senapati, Z. H. Barber, and M. G. Blamire, Adv. Mater. **25**, 5581-5585 (2013).

Simulation of the temperature profile of $BaCaZrTiO_3$ thin films during laser annealing

Tiago Rebelo[1,*], *João* Alves[1], and *Bernardo* Almeida[1]

[1]Centre of Physics/Department of Physics, University of Minho, Gualtar Campus, Braga, Portugal

Abstract. The laser annealing of a $Ba_{0.85}Ca_{0.15}Ti_{0.9}Zr_{0.1}O_3$ (BCZT) thin film on a metglas substrate was simulated in order to understand how the annealing parameters (energy and fluence of the laser, pulse duration, etc) influence the optimization of the crystallinity of the film. Using a 1D heat diffusion equation combined with a finite difference method, the variation of the temperature with the depth relative to the film's surface and on annealing time was studied. The laser intensity, BCZT's reflectivity and the temperature dependence of the thermal conductivity and specific heat of the BCZT were considered. No structural phase changes were detected in both the BCZT and the metglas for the values of laser fluence studied, but for 80 mJ/cm^2 the maximum temperature approached near the BCZT's melting temperature. It was observed that since the film's thermal conductivity decreases with increasing fluence, lower fluences allow for a better distribution of the laser's energy throughout the crystal lattice, increasing the crystallinity. It was further observed that between consecutive pulses the film's temperature stabilizes at room temperature.

1 Introduction

Multiferroic piezoelectric-magnetostrictive composites have been attracting much attention, since they allow exploring coupled electric/magnetic responses [1, 2]. Composite nanostructures made of $Ba_{0.85}Ca_{0.15}Ti_{0.9}Zr_{0.1}O_3$ (BCZT, ferroelectric) thin films on Metglas 2826MB (ferromagnetic) substrates are one of such composites, combining the high piezoelectricity of the BCZT with the high magnetostriction of the metglas, originating a multiferroic material with potential enhanced coupled responses. However, when producing the thin film, high deposition temperatures may be required [2] that modify the substrate properties. In such cases a possible approach is to perform the deposition at low temperatures followed by an annealing process using a laser with a low fluence (*laser annealing*), which will remove defects that may have accumulated during the deposition, promote the film crystallization and avoid heating the substrate [3]. This annealing process is particularly indicated when the materials involved can't withstand the temperatures reached in a conventional annealing in an oven, as is the case with the materials used in this work: the metglas substrate has an irreversible structural phase change (crystallization) at 450 °C that degrades its magnetic properties, while the BCZT film requires a deposition temperature of approximately 700 °C. By depositing the BCZT layer at 400 °C over the metglas substrate and subsequently laser annealing the film, the crystallization of the BCZT layer can be achieved without affecting

*e-mail: rebelo.tiago@gmail.com

the metglas' properties. There is then interest in studying how the film's temperature varies during this process, both with the depth relative to its surface and with time, as it allows the optimization of the annealing conditions (laser energy and fluence, pulse duration, etc). To this end, the variation of the laser pulse intensity with time and the temperature dependence of the specific heat and thermal conductivity have to be considered. The effect of the temperature on the substrate is also of interest. It was observed that the laser does not contribute appreciably to its heating, as envisaged. The main effect of the substrate in the composite is the absorption of energy from the film and, consequently, the lowering of its temperature.

Given the similarity between BCZT and barium titanate ($BaTiO_3$), its base compound, the properties of the later can be used instead of those of the former in the calculations. This was done during this study whenever the desired BCZT properties could not be found.

2 Experimental procedure

Given that the focus of this work is the modeling of the temperature profile during the laser annealing, only a brief description will be given reagarding the film growth technique. More detailed descriptions can be found in [4, 5]. The film was grown via pulsed laser ablation with a KrF Lambda Physik LPXPro 210 laser with wavelength $\lambda = 248$ nm and a pulse duration of 18 nanoseconds (FWHM). The same laser was used in the laser annealing procedure. The BCZT film was deposited at 400 °C under an oxygen atmosphere, with a pressure of 0.02 mbar, and a fluence of 2 J/cm^2. To perform the laser annealing process, the film was struck with the same laser but using a lower energy than that used during the deposition. Also, the beam was defocused to reduce the fluence. A frequency of 10 Hz was used and 3 values of laser fluence were studied: 40, 50 and 80 mJ/cm^2. The annealing process was performed during 200 seconds in air, to preserve the film's oxidation level. Figure 1 shows the experimental apparatus for the procedure.

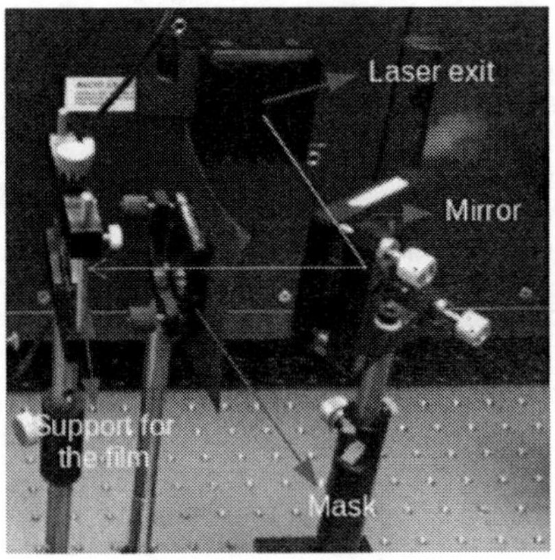

Figure 1. Experimental apparatus for the laser annealing process. The laser path is shown in green.

3 Heat diffusion in the samples

The geometry of the film, as well as its positioning relative to the laser during the annealing process, allow the heat diffusion to be well described by the 1D Fourier heat diffusion equation [3]:

$$\rho_f C_P \frac{\partial T(x,t)}{\partial t} - \frac{\partial}{\partial x}\left[k\frac{\partial T(x,t)}{\partial x}\right] = P(x,t) \tag{1}$$

where $T(x,t)$ is the temperature at depth x relative to the film's surface and at time t, ρ_f is the density of the film [6], C_P and k are its specific heat and thermal conductivity respectively, and $P(x,t)$ is the heat generation function, a function that models the effect of the heat source - in this case, the laser. However, when C_P and k vary strongly with temperature, as in our laser heated samples, equation 1 can't be directly used to obtain an analytical solution [3].

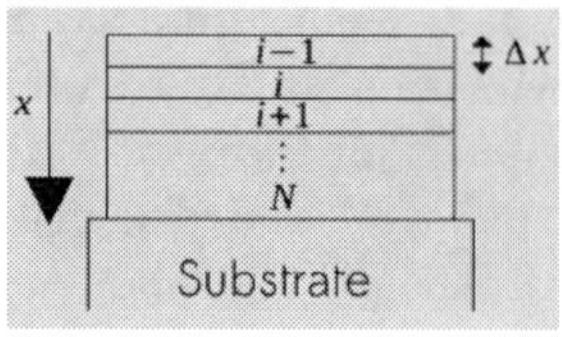

Figure 2. Division of the film in several layers of equal thickness during the discretization procedure.

An alternative is to discretize the problem: the film is divided in several layers of equal thickness (Figure 2) originating an equation for each layer.

We can then obtain the temperature of each layer, i, at the next instant $(n+1)$ based on the temperature of that layer at the previous instant (n), as well as the temperatures and thermal conductivities of the adjacent layers at the previous instant (equation 2). The substrate can be included in the model by considering it as an extension of the film and applying the discretization procedure. In the film-substrate interface, however, the density, thermal conductivity and specific heat of the metglas have to be considered.

$$T_i^{n+1} = T_i^n + \frac{\Delta t}{\rho_f C_{P_i}^n} P(x,t) + \frac{\Delta}{\Delta x^2} \frac{1}{\rho_f C_{P_i}^n} \left[k_{i-1}^n \left(T_{i-1}^n - T_i^n \right) + k_{i+1}^n \left(T_{i+1}^n - T_i^n \right) \right] \tag{2}$$

We also need boundary conditions. The chosen conditions were: 1) there is no heat loss through the surface [3], which means that the surface layer is at the same temperature as the atmosphere immediately in contact with it - $T_0^n = T_{-1}^n$; 2) the sample is thick enough for the temperature to stabilize with increasing distance from the surface, which means that the final layer of the substrate (layer N) is in thermal equilibrium with the following layer - $T_N^n = T_{N+1}^n$. The following layer would be the sample support, which is at room temperature, so this boundary condition is simply $T_{N+1}^n = 293.15$ K. Applying these conditions to equation 2 we obtain a set of equations for each layer of the film and substrate:

$i = 0$:

$$T_i^{n+1} = T_i^n + \frac{\Delta t}{\rho_i C_{P_i}^n} P_i(x,t) + \frac{\Delta t}{\Delta x^2} \frac{1}{\rho_i C_{P_i}^n} k_{i+1}^n \left(T_1^n - T_0^n \right) \tag{3}$$

$i = 1, ..., N-1$:

$$T_i^{n+1} = T_i^n + \frac{\Delta t}{\rho_i C_{P_i}^n} P_i(x,t) + \frac{\Delta t}{\Delta x^2} \frac{1}{\rho_i C_{P_i}^n} \left[k_{i-1}^n \left(T_{i-1}^n - T_i^n \right) + k_{i+1}^n \left(T_{i+1}^n - T_i^n \right) \right] \tag{4}$$

$i = N$:

$$T_i^{n+1} = T_i^n + \frac{\Delta t}{\rho_m C_{P_m}} P_m(x,t) + \frac{\Delta t}{\Delta x^2} \frac{1}{\rho_m C_{P_m}} k_m \left[\left(T_{i-1}^n - T_i^n \right) + (293.15 - T_i^n) \right] \tag{5}$$

where ρ_i, C_{P_i} and $P_i(x,t)$ are, respectively, the density, specific heat and heat generation function of the material that constitutes the ith layer, and ρ_m, C_{P_m}, k_m and $P_m(x,t)$ are, respectively, the density [7], specific heat ([8], refering to Metglas 2605S-2, but expected to be similar), thermal conductivity [9] and heat generation function of the metglas. The values of the ith layer's physical properties correspond to either BCZT or metglas, depending on the position.

Finally, for the simulations performed, a film thickness of approximately 500 nm and substrate thickness of 10 μm were chosen as reasonable values, in accordance with the scanning electron microscopy studies performed on the samples. After some experimentation, the time step $\Delta t = 1$ ns and layer thickness $\Delta x = 65$ nm were chosen as the smallest representative values that the computational resources available could handle.

The heat generation function is of the form $P(x,t) = C(1-R)I(t)\exp(-\alpha x)$, where C is a normalization constant, R and α are, respectively, the reflectivity and absorption coefficient of the BCZT, $I(t)$ is the time dependent laser profile pulse intensity and x is the depth relative to the film's surface. The reflectivity of the BCZT can be determined considering that the laser light is unpolarized and using barium titanate's refractive index ([10], $\lambda = 244.3$ nm).

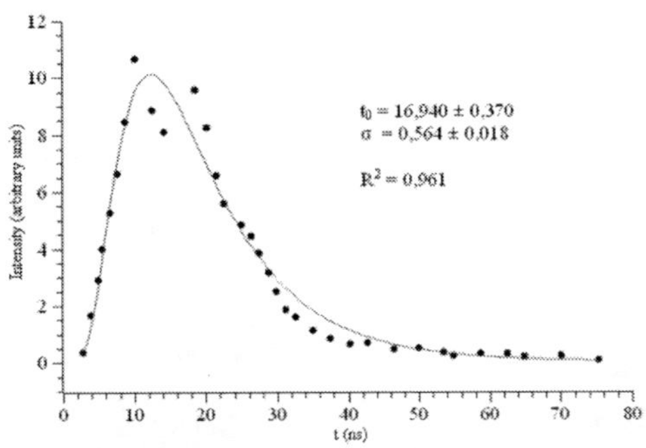

Figure 3. Fitting of the laser profile with a lognormal dependence on time. t_0 is related to the instant at which $I(t)$ is maximum and σ is the standard deviation. R^2 is the statistical measure of goodness of fit.

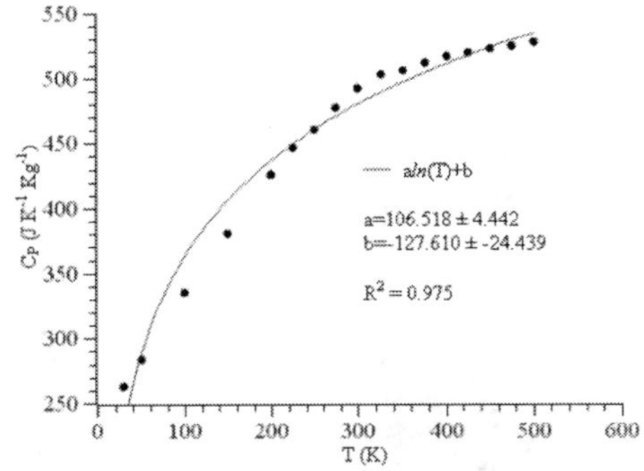

Figure 4. The specific heat follows reasonably well a logarithmic dependence on the temperature (data from [11].

The absorption coefficient is related to λ and to the extinction coefficient of the BCZT (taken from [10], for $BaTiO_3$). In regard to $I(t)$, the laser profile was obtained from the manufacturer and it was fitted (Figure 3) with a lognormal function. It is observed that $I(t) \approx 0$ from 50 ns onwards. So, this value was taken as the pulse duration, τ. Attending to the normalization condition for $P(x,t)$, $C = F_0/\tau$, where F_0 is the fluence of the laser, we finally obtain the heat generation function for the BCZT:

$$P(x,t) = (1-R)\frac{F_0}{\tau}\frac{1}{(t/t_0)\sigma\sqrt{2\pi}}\exp\left(\frac{-\ln^2(t/t_0)}{2\sigma^2}\right)e^{-\alpha x} \qquad (6)$$

The used t_0 and σ values are in Figure 3. The heat generation function for the substrate depends on the amount of laser radiation that reaches its surface layer, that is, it depends on the value of $P(x,t)$ at the final layer of the BCZT. Our calculations show that this value is < 0.05 W m^{-3}, so we conclude that $P_m(x,t) \approx 0$. Consequently, the second term on the right hand side of equations 4 and 5 becomes negligible in the substrate case and was dropped from its calculations.

The temperature dependence for the BCZT's specific heat, $C_P(T)$, was obtained by fitting a curve to experimental data (Figure 2(b) of [11]). A logarithmic dependence on temperature was fitted (Figure 4). This is reasonable from a physical point of view, as it is expected that $C_P \to$ constant as $T \to \infty$.

For the temperature dependence of the thermal conductivity, $k(T)$, there is indication [12] that, for $BaTiO_3$, it decreases with increasing temperature more slowly than $k \propto T^{-1}$ in the temperature range here considered. The expected dependence will thus be of the form:

$$k(T) = \frac{A}{T^\beta}, \quad \beta < 1 \qquad (7)$$

where A and β are constants determined by attending to the boundary conditions given in [12], $k(T) = 2.51$ W m^{-1}K^{-1} for $T = 1000$ K and $k(T) = 4.99$ W m^{-1}K^{-1} for $T = 300$ K, but adapting the second boundary condition to the fact of the BCZT being a ceramic, so that $k(300) = 2.85$ W m^{-1}K^{-1} [6]. This resulted in $A \approx 73.9$ W m^{-1} and $\beta \approx 0.571$.

 https://doi.org/10.1051/epjconf/202023305008

4 Results

The temperature profiles $T(x,t)$ for the different values of laser fluence studied (40, 50 and 80 mJ/cm^2) are presented in Figures 5, 6 and 7. Only the layers of the substrate that exhibited some heating are shown.

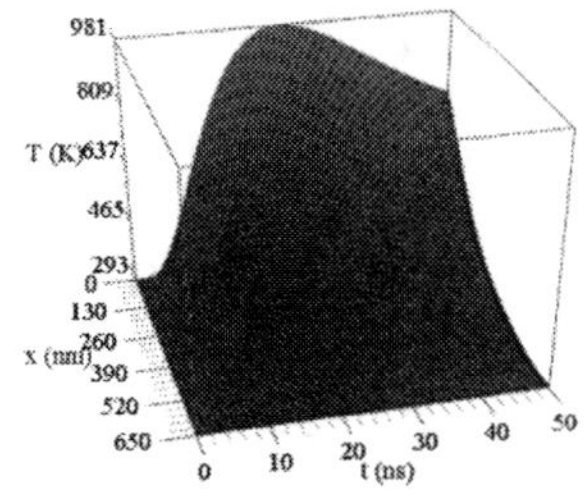

Figure 5. Temperature profile $T(x,t)$ for $F_0 = 40$ mJ/cm^2.

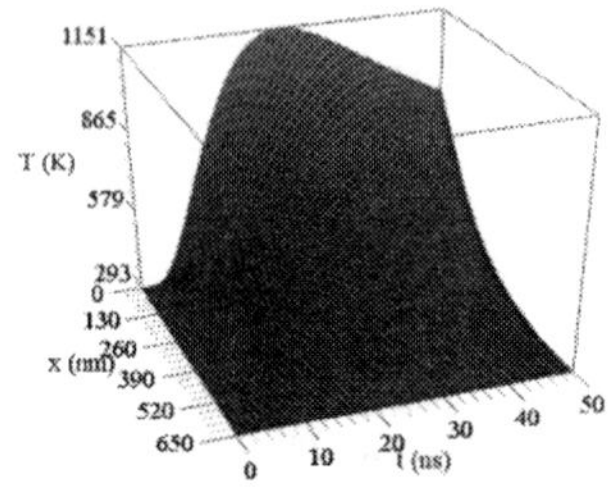

Figure 6. Temperature profile $T(x,t)$ for $F_0 = 50$ mJ/cm^2.

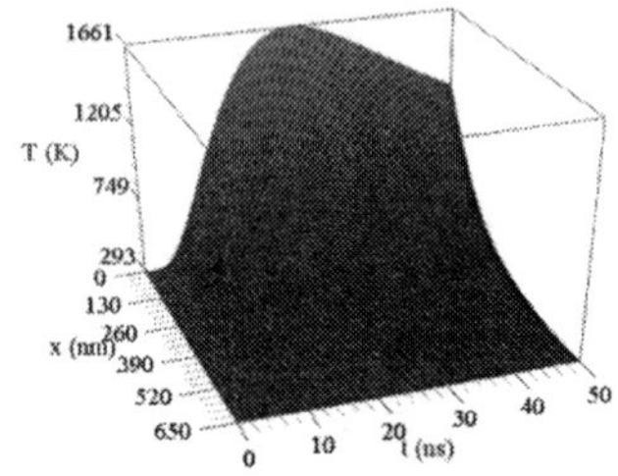

Figure 7. Temperature profile $T(x,t)$ for $F_0 = 80$ mJ/cm^2.

As expected, an increase in laser fluence increases the reached temperatures, but no melting of the BCZT is observed ($BaTiO_3$ melting temperature: $\approx$1898 K [13]). The temperatures reached in the highest fluence studied, however, approach the melting temperature, but only on the most superficial layer of the BCZT. This is expected from the behaviour $k \propto T^{-1}$ of the thermal conductivity. As the temperature rises due to higher fluences, k will decrease in the considered temperature range, which means that the laser energy won't be as easily transmitted to the inner layers of the film. This leads to an accumulation of energy on the superficial layer and considerable increase of its temperature in comparison with the rest of the film. As such, lower fluences may be more desirable during the annealing process, as they will generate lower temperatures and, consequently, won't decrease the thermal conductivity. This will promote the propagation of the laser energy to the inner layers of the film, originating a more uniform distribution of the energy and promoting the film crystallinity, the intended goal of the annealing process.

Regarding the substrate, it was observed that only its most superficial layers experience some small heating. However, there is interest in stuying how the temperature evolves between consecutive pulses, since the propagation of energy to the substrate may increase the temperature further during this interval. Since each pulse has an effective duration $\tau = 50$ ns and the laser frequency is 10 Hz, the time between consecutive pulses is about 10^8 ns.

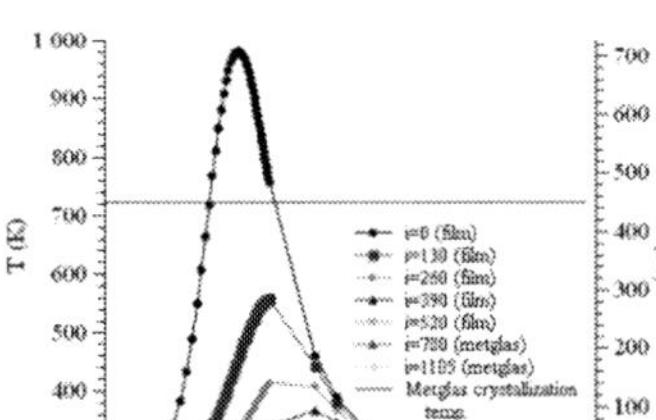

Figure 8. Temperature profile $T(t)$ for $F_0 = 40$ mJ/cm^2.

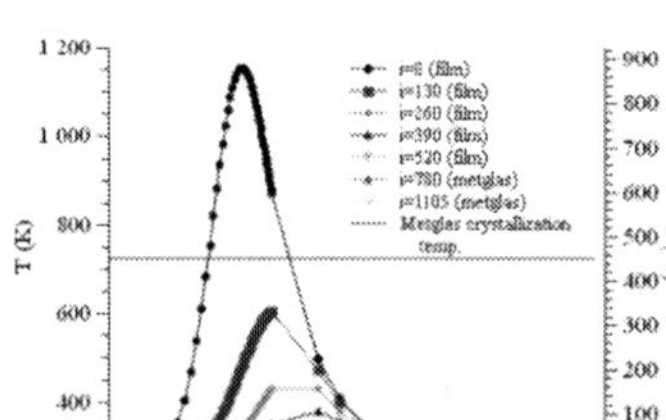

Figure 9. Temperature profile $T(t)$ for $F_0 = 50$ mJ/cm^2.

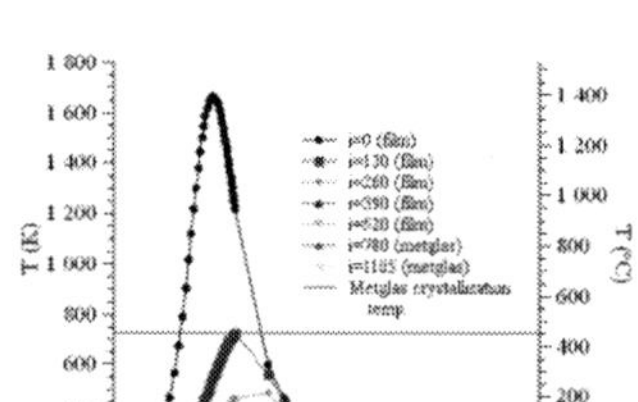

Figure 10. Temperature profile $T(t)$ for $F_0 = 80$ mJ/cm^2.

Figures 8, 9 and 10 show the simulations of the variation of the temperature with time, $T(t)$, for several reference layers in both film and substrate. The metglas' crystallization temperature is also represented (pink horizontal line). No fluence induces crystallization in the metglas. It is also observed that all layers, both from the BCZT and the metglas, stabilize at room temperature before the beginning of the next pulse, independently of the fluence. As such, from a thermal viewpoint the situation is the same in each pulse, always reaching the same level of heating in each layer. This suggests that an extension of the laser pulse in time, in such a way that between consecutive pulses no temperature stabilization occurs (but without provoking crystallization of the metglas), would allow for energy accumulation in the film. This, combined with a low laser fluence, would provide an uniform energy distribution throughout the crystal lattice, optimizing the film crystallinity without affecting the metglas.

5 Conclusions

We developed a model to simulate the temperature profile $T(x,t)$ of BCZT thin films on metglas substrates, during a laser annealing process. The model involved the construction of a heat generation function (describing the effect of the laser on the sample), as well as taking into account the temperature dependence for the specific heat and the thermal conductivity. The temperature profiles obtained revealed that no melting of the BCZT occurs for any of the fluences studied. Further, it revealed that the thermal conductivity plays a relevant role in the temperatures achieved and layers affected by the heating, indicating that lower fluences may be more desirable to promote the uniform distribution of energy throughout the crystal lattice and, consequently, the crystallinity of the film. It was observed that the metglas doesn't experience significant heating, as envisaged, and that between consecutive pulses the temperature of both film and substrate stabilizes at room temperature.

References

[1] N.A. Spaldin & R. Ramesh, Nature Materials **18**, 203-212 (2019)

[2] J.G. Barbosa, I.T. Gomes, M.R. Pereira, C. Moura, J.A. Mendes and B.G. Almeida, J. Appl. Phys. **116**, 164112 (2014)

[3] R.F. Wood & G.E. Giles, Physical Review B **23**, 2923-2942 (1981)

[4] C. Araújo, B.G. Almeida, M. Aguiar, J.A. Mendes, Vacuum **82**, 1437-1440 (2008)

[5] I.T. Gomes, B.G. Almeida, A.M.L. Lopes, J.P. Araújo, J. Barbosa, J.A. Mendes, J. Magn. Magn Mater. **322**, 1174-1177 (2010)

[6] Yi He, Thermochimica Acta **419**, 135-141 (2004)

[7] Metglas Inc., *Magnetic Alloy 2826MB (nickel-based) Technical Bulletin* (Metglas Inc., 2011) 1

[8] R.J.J. Martis, T.J. Taylor and Y.P. Khanna, Journal of Non-Crystalline Solids **94**, 209-215 (1987)

[9] Metglas Inc., *https://metglas.com/frequently-asked-questions/\#question_twentyeight* (Metglas Inc., 2017)

[10] E. Palik (edited by), *Handbook of optical constants of solids II* (Academic Press, Inc, 1991)

[11] R. Choithrani, Invertis Journal of Science and Technology **7**, 72-77 (2014)

[12] A. van Roekeghem, J. Carrete, C. Oses, S. Curtarolo, N. Mingo, Physical Review X **6**, 041061 (2016)

[13] D.R. Lide, *CRC Handbook of Chemistry and Physics, 89th Edition (Internet Version 2009)* (CRC Press/Taylor and Francis, 2009) 4-51

On the Single Wall Carbon Nanotube Surface Plasmon Stability

Jaime Silva[1,*], *Bruce* F. Milne[1,**], and *Fernando* Nogueira[1,***]

[1]CFisUC, Department of Physics, University of Coimbra, Rua Larga, 3004-516 Coimbra, Portugal.

Abstract. The physics of surface plasmons has a long tradition in condensed matter theory but as the dimension of the systems reaches the nano scale, new effects appear. In this work, by calculating the absorption spectra of a single wall carbon nanotube, using time dependent density functional theory, the effect of adding/removing electrons on the surface plasmon energy is studied. It is shown that removing electrons from the single wall carbon nanotube does not affect the surface plasmon energy peak. In contrast, adding electrons to the single wall carbon nanotube will redshift the plasmonic peak energy, an effect that is explained by an increase of the electron effective mass.

1 Introduction

The surface plasmon on metals has a long history that started with the explanation of experimental results by Pines and Bohm [1–3]. Pines and Bohm explained that the longitudinal oscillation of the electron density observed in thin metal films when bombarded with fast electrons is due to long-range coulomb interaction of the valence electrons in the metal films [1–3]. The latter authors coined the term "plasmon" to describe the collective oscillation of the electron density due to the similarity with the electronic plasma oscillations in gas discharges [1–3]. Finally, Stern and Ferrell [4] also described these collective oscillations of the density as surface plasmons.

It is important to distinguish between bulk and surface plasmons. Using classic electrodynamics the bulk plasmon dispersion relation can be retrieved by solving $\nabla \cdot \mathbf{D} = 0$ for field propagating in a medium with no sources, giving two solutions. The first solution of the latter equation is the bulk plasmon and second is the dispersion relation for photons propagating in a medium which is called a polariton [1, 5]. The surface plasmon dispersion relation is obtained following a similar procedure but solving the $\nabla \cdot \mathbf{D} = 0$ for electric fields propagating in the upper and lower half-spaces of the surface.

As the dimension of the system is reduced, a quantum description of the surface plasmon is needed. The field that studies the plasmonic oscillations using quantum mechanics is termed quantum plasmonics [6]. The use of quantum mechanics to describe the surface plasmons comes from the fact that classical electrodynamics fails to describe plasmonic resonances at sub-nanometer gaps [7]. In fact, Drude's model is usually employed to describe

*e-mail: jaime.d.silva@uc.pt
**e-mail: bfmilne@uc.pt
***e-mail: fnog@uc.pt

the surface plasmons of small nanoparticles. Nevertheless, Drude's model fails in the prediction of the blueshift of the plasmonic peak energy when decreasing the nanoparticle size [8]. The latter is due to the fact that for small particles the dielectric function cannot describe a small molecule. On way to overcome the limitations of the classic approach is to use time dependent density functional theory (TDDFT) [9, 10]. Using TDDFT, Zuloaga et. al. [11] found that the optical spectra is very different from the ones obtained using classical calculations. The latter authors concluded that the electron spill-out at the surface induces a screening that is not captured using a classical calculation. Townsend et. al. [12], using TDDFT, characterised the resonances of metallic nanoparticles investigating which characteristics are important to decide if a resonance is plasmonic. In another related work, Yan et. al. [13], using TDDFT, demonstrated for a linear atomic chain the existence of end and central plasmon resonances. The existence of plasmon resonances at the nanoscale opens the route for plasmonic nanosensors. In fact, as discussed in [14], this type of sensors are able to detect single molecules. Furthermore, the existence of surface plasmon resonance modes in nanoparticles enables their use as biochemical sensors, among other applications [14]. The observation of a surface plasmon in the optical response of single-wall nanotubes (SWCNT) [15] can pave the way for the construction of new type SWCNT plasmonic sensors with a vast application range. In a previous work [16] the plasmonic peak in the optical spectra of SWCNT was studied in detail. The purpose of this work is to study, theoretically, the stability of the surface plasmon seen in metallic carbon nanotubes [16] upon electron addition, removal and adsorption of organic gases.

This work is organized as follows. In the methodology section we present the details of the theoretical calculation. This is followed by the results and discussion section and the main conclusions.

2 Methodology

In this work we use density functional theory (DFT)[17] and TDDFT to calculate[18] the absorption spectra, the paramagnetic current and the density of states (DOS) of a (3,3) SWCNT with removed/added electrons. In the framework of DFT/TDDFT the Hamiltonian is given by Equation 1.

$$\hat{H} = \frac{1}{2}(-i\nabla + \frac{1}{c}\mathbf{A}_{ext})^2 + \hat{v} \tag{1}$$

where $\mathbf{A}_{ext}$ is the external vector potential. For an uniform static magnetic field $\mathbf{B}$ applied to a finite system with zero total spin, the vector potential, $\mathbf{A}$, can be defined using the Coulomb gauge by:

$$\mathbf{A}(\mathbf{r}) = \frac{1}{2}\mathbf{B} \times \mathbf{r} \tag{2}$$

The time-dependent Kohn-Sham (KS) orbitals are determined using the Hamiltonian of Equation 1 where $\hat{v}$ is the KS effective potential that can be decomposed in an external potential, an Hartree potential and an exchange-correlation potential. In our calculations we used a uniform real space grid to discretise the equations. Spacing was 0.20 Å and the simulation box was constructed by joining spheres of radius 3.5 Å around every atom. For the time domain, a time step of 0.000658 fs was employed for the propagation of the KS equations. Finally, the exchange-correlation potential was approximated by using the local density approximation.

In this work, the finite SWCNT were generated [1] with 15 translations of the unit cell along the tube orientation and the terminations of the finite SWCNT were passivated with hydrogens. Finally, all the structures were optimised within DFT using the exchange functional of Tao et el (TPSS) [19] and using Def2-TZVP basis sets with Grimme's D3 dispersion corrections, as provided by the ORCA package[20].

3 Results and discussion

As the size of the system diminishes, the stability of the surface plasmon upon electron removal and addition becomes increasingly important for the final electronic design of the device. In Figure 1 we present the theoretical spectra of a SWCNT when pairs of electrons are removed from it.

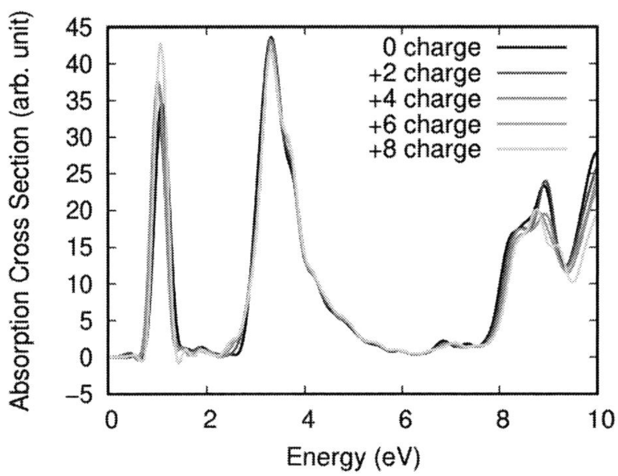

Figure 1. Absorption spectra of a SWCNT when electrons are removed from it.

It is possible to observe in Figure 1 that the plasmonic peak[16] energy, the first one, remains the same when the electrons are removed. In fact, by observing the density of states (DOS) of SWCNT, Figure 3, it can be shown that it remains the same. But absorption spectra peak heights depend on the number of electrons in each system.

In Figure 2 we present the absorption spectra for a (3,3) SWCNT with added electrons. As can be seen in the latter figure, adding electrons to the SWCNT will redshift the surface plasmon peak.

This redshift can be understood by looking at the DOS presented in Figure 4, where it can be seen that adding electrons to a SWCNT will redshift the DOS. Hence, this shift in DOS will induce the observed shift in the absorption spectra. Interestingly, the DOS does not change when electrons are removed, as can be seen in Figure 3.

A possible explanation for the change in the DOS upon electron addition is the increase of the effective mass of the electrons[21] due to Coulomb repulsion, that will reduce the excitation energy for low bounded electrons. It should be stressed that in phenomenological models of [21] the plasmonic regime, a critical density that depends on the motion of electrons dictated by Coulomb repulsion rather than a confinement of the carriers is achieved. It is

[1]TubeGen 3.4 (web-interface, http://turin.nss.udel.edu/research/tubegenonline.html), J. T. Frey and D. J. Doren, University of Delaware, Newark DE, 2011

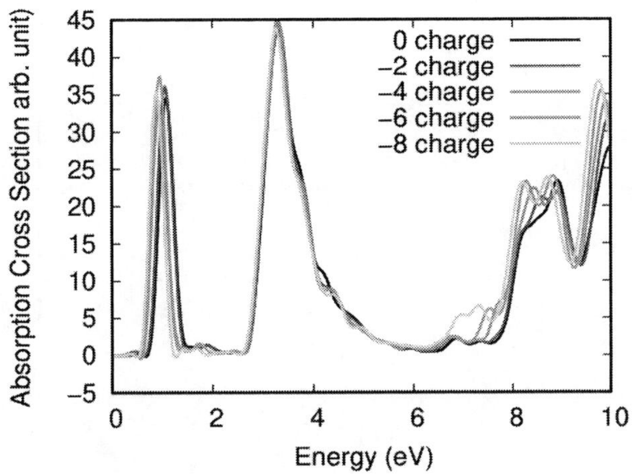

Figure 2. Absorption spectra of a SWCNT when electrons are added to it.

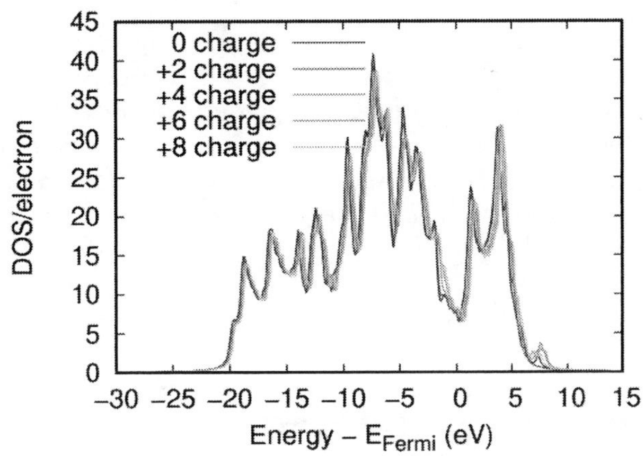

Figure 3. DOS for a SWCNT when electrons are removed from it.

possible to observe in their work a redshift of the plasmon peak with the increase of the carrier effective mass. One point that should be discussed regarding the model of [21] is that, for the latter model, adding carriers to a system will blueshift the energy of the surface plasmon, a behaviour contrary to what is observed in this work. This can be explained by the high electron density of the metallic SWCNT, that remains practically unchanged with the addition of a few electrons.

Taking in mind the last observation, we concluded that the electrons added to the SWCNT will not participate directly in the collective resonance that forms the surface plasmon, but will contribute to an increase of the SWCNT electron effective mass that, as discussed in [21], will redshift the surface plasmon energy.

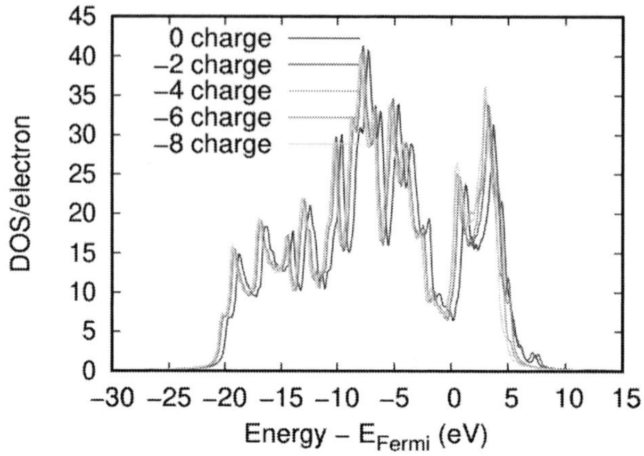

Figure 4. DOS for a SWCNT when electrons are added to it.

4 Conclusion

In this work we study the stability of the surface plasmon upon electron removal/addition. We demonstrate that removing electrons from a (3,3) SWCNT does not affect the plasmonic peak energy present in the SWCNT absorption spectrum. But electron addition will induce a redshift in the plasmonic peak energy, correlated to a redshift of the DOS. This redshift of the plasmonic energy peak with the addition of electrons can be related to the increase of the electron effective mass that, due to Coulomb repulsion, reduces the excitation energy for low bounded electrons.

5 Acknowledgments

We thank the Laboratory for Advanced Computing (LCA) of the University of Coimbra for the computer time allocated to this project.

References

[1] I.R. Hooper, W.L. Barnes, in *Handbook of Surface Science* (Elsevier B.V., 2014), Vol. 4, pp. 34–74
[2] D. Pines, D. Bohm, Phys. Rev. **85**, 338 (1952)
[3] D. Pines, Rev. Mod. Phys. **28**, 184 (1956)
[4] E.A. Stern, R.A. Ferrell, Phys. Rev. **120**, 130 (1960)
[5] J.J. Hopfield, Phys. Rev. **112**, 1555 (1958)
[6] M.S. Tame, K.R. McEnery, S.K. Ozdemir, J. Lee, S.A. Maier, M.S. Kim, Nat. Phys. **9**, 329 (2013)
[7] R. Esteban, A.G. Borisov, P. Nordlander, J. Aizpurua, Nat. Commun. **3**, 825 (2012)
[8] E.B. Guidez, C.M. Aikens, Nanoscale **6**, 11512 (2014)
[9] E. Runge, E.K.U. Gross, Phys. Rev. Lett. **52**, 997 (1984)
[10] M. Marques, N. Maitra, F. Nogueira, E.K.U. Gross, A. Rubio, *Fundamentals of Time-Dependent Density Functional Theory*, Lecture Notes in Physics (Springer Berlin Heidelberg, 2012)

[11] J. Zuloaga, E. Prodan, P. Nordlander, ACS Nano **4**, 5269 (2010)
[12] E. Townsend, G.W. Bryant, J. Optics-Uk **16**, 114022 (2014)
[13] J. Yan, Z. Yuan, S. Gao, Phys. Rev. Lett. **98**, 216602 (2007)
[14] J.N. Anker, W.P. Hall, O. Lyandres, N.C. Shah, J. Zhao, R.P. Van Duyne, Nat. Mater. **7**, 442 (2008)
[15] T. Morimoto, S.K. Joung, T. Saito, D.N. Futaba, K. Hata, T. Okazaki, ACS Nano **8**, 9897 (2014)
[16] J. Silva, M.J.T. Oliveira, S. Lanceros-Mendez, F. Nogueira, J. Phys. Chem. C **120**, 18268 (2016)
[17] P. Hohenberg, W. Kohn, Phys. Rev. **136**, B864 (1964)
[18] X. Andrade, D. Strubbe, U. De Giovannini, A.H. Larsen, M.J.T. Oliveira, J. Alberdi-Rodriguez, A. Varas, I. Theophilou, N. Helbig, M.J. Verstraete et al., Phys. Chem. Chem. Phys. **17**, 31371 (2015)
[19] J. Tao, J.P. Perdew, V.N. Staroverov, G.E. Scuseria, Phys. Rev. Lett. **91**, 146401 (2003)
[20] F. Neese, Wires Comput. Mol. Sci. **2**, 73 (2012)
[21] P.K. Jain, J. Phys. Chem. Lett. **5**, 3112 (2014)

Numerical Simulation of Non-Equilibrium Stationary Currents Through Nano-scale One-Dimensional Chains

João Pedro dos Santos Pires[1,2,*], *Bruno* Amorim[1,3], and *João Manuel* Viana Parente Lopes[1,2]

[1]Centro de Física das Universidades do Porto e Minho
[2]Faculty of Sciences, University of Porto, Rua do Campo Alegre 4169-007, Porto, Portugal
[3]University of Minho, Campus of Gualtar, 4710-057, Braga, Portugal

Abstract. Using a method based on the time-evolution of the occupied states at zero temperature, we observe the onset of a quasi-uniform and quasisteady state current across a disordered tight-binding chain, coupled between two finite (but large) clean leads with open boundaries. This current is seen to match the one obtained in the Landauer-Büttiker formalism and is also independent of the initial condition considered (partitioned or non-partitioned). Finite-size effects are also reported and briefly discussed.

1 Introduction

Electron transport in solid-state systems has always been an important topic in condensed matter theory due to both its technological importance and the interesting quantum-mechanical effects it displays. In bulk materials, electron transport due to a static electric field is usually described in by a local conductivity, σ_{DC}. From the theoretical side, this local conductivity is computed using different Kubo formulas (KF) [6], for which several analytical and very efficient numerical methods have been developed [1, 4, 5, 10, 16].

With the advent of nano-scale devices in the 1980's, the need for a different approach to this problem emerged. At low temperatures and for tiny samples, electrons propagate as coherent matter waves which are highly affected by the specific distribution of impurities and the presence/geometry of the electric contacts used in the experiment. In such situations, describing transport by a local conductivity is inadequate (see Ref. [14] for details). These issues were eventually settled by the application of a different formalism: the Landauer-Büttiker formalism (LBF).

The LBF [2, 7] recasts quantum electron transport as a wave scattering problem. The current that flows across a sample is described in terms of the transmission of incoming waves that are partially reflected and partially transmitted by the sample. These incoming waves originate from perfect and semi-infinite leads, which simulate the electrodes in an experiment. It is also assumed that the occupation of these propagating states is controlled by the distinct chemical potential of each lead. In the LBF, the current through the sample, at zero temperature, is given by the *Landauer Formula* [8, 17]:

$$I_{\text{DC}} = \frac{e}{h}\int_{E_F-e\frac{\Delta V}{2}}^{E_F+e\frac{\Delta V}{2}} dE\,\mathcal{T}(E), \tag{1}$$

*e-mail: up201201453@fc.up.pt

where $\mathcal{T}(E)$ is the transmission function, $E_F \pm \Delta V/2$ is the chemical potential of each lead, with ΔV the potential bias that is applied between the two leads and E_F a reference chemical potential. Despite its simple appearance, the Eq. (1) describes a fully non-linear DC current, where the integrand can be identified with the quantum transmittance, $\mathcal{T}$, of the sample at energy E. For small bias, the Landauer Formula reduces to $I_{\rm DC} = G_{\rm lin}\Delta V$, where $G_{\rm lin}$ is the linear conductance, which is a function of E_F, and is proportional to the transmission:

$$G_{lin}(E_F) = \frac{e^2}{h}\mathcal{T}(E_F). \tag{2}$$

It must be pointed out that the Landauer formula only describes the steady-state current and requires the assumption that the leads have well-defined chemical potentials, which are ill defined out of equilibrium.

The issue of how this steady-state is achieved and if it even exists has been a subject of inquiry over the years [3, 13]. When trying to address this problem, the question of what is the initial condition of the system, becomes relevant. It is therefore natural to consider two different initial conditions:

- ***Partitioned Setup***: In which the leads and sample are initially disconnected, each being in a local equilibrium state described by a chemical potential and subject to a potential bias. The leads and sample are then brought suddenly into contact;
- ***Non-Partitioned Setup***: A more realistic scenario, in which the leads and sample are initially connected and in equilibrium, with a common chemical potential. Then a bias potential between the two leads is abruptly applied.

It has been shown that in the long time limit, and provided the leads are infinite (more rigorously, provided they form a continuum of states), a steady-state is achieved, independent of the initial condition, with a current described by the Landauer Formula [13].

However, the current dynamics for finite leads has never been investigated. Such is the goal of this work. Hence, we study the time-evolution of the current in a prototypical 1D system, propagating in time the occupied quantum states of the system. Both *partitioned* and *non-partitioned* initial conditions are used and a comparison with the steady-state Landauer current is made. In both cases, the current was measured in different places of the central sample, for different disorder profiles and increasing lead sizes.

The paper is organized as follows: In Sect. 2 we introduce the model and essential definitions used for our numerical studies. In Sect. 3, we outline the basics of the method used to calculate the time-dependent current in this work and present our results. Finally, in Sect. 4, discuss our results and present conclusions.

2 Model Hamiltonian and Local Current

In order to study the current dynamics and the possible emergence of steady-state transport in a finite system, we will consider the prototypical one-dimensional tight-binding model , with nearest neighbor hoppings, represented in Fig. 1. This is described by the Hamiltonian:

$$\mathcal{H}(t) = \sum_{n=0}^{L-1}\left(\epsilon_n^{\rm d} + v_n^{\rm ext}(t)\right)c_n^\dagger c_n - \sum_{n=0}^{L-2} w_n^{n+1}(t)\left(c_{n+1}^\dagger c_n + c_n^\dagger c_{n+1}\right). \tag{3}$$

The chain has L sites, which we split into left and right leads, and a central sample. As depicted in Fig. 1, the left (right) lead, is formed by L_l sites - $0 \le n \le L_l{-}1$ ($L_l{+}Ls \le n \le L{-}1$)

- and the central sample is formed by L_s sites - $L_l \leq n \leq L_l + L_s - 1$. In will consider the limit of very large leads ($L_l \gg L_s$).

Sites in the central sample are subject to an on-site random disorder potential $\epsilon_n^{\rm d}$, uniformly sampled from $\left[-\frac{W}{2}, \frac{W}{2}\right]$. $w_n^{n+1}(t)$ are the nearest-neighbors hoppings with constant value w, except for the sample-lead bonds, and $v_n^{\rm ext}(t)$ is an external bias potential, constant inside the leads (no electric field) and linear in the sample region (uniform electric field), as illustrated in Fig. 1. For $t > 0$, the potential profile is given by V_n, where:

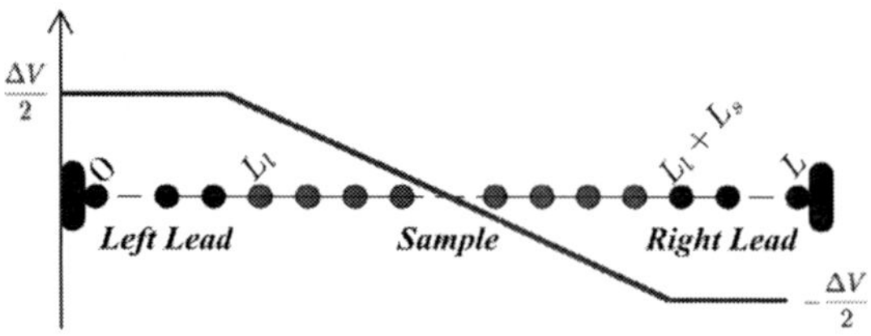

Figure 1. Prototype model used to simulate transport across an one-dimensional sample coupled to finite leads. The red dots are the places where there is a disordered potential and the blue curve is the profile of the externally applied potential. The chain has open boundaries.

$$V_n = \mp\tfrac{e}{2}\Delta V \quad \text{if } n \in \text{Left (Right) Lead} \quad \text{and} \quad V_n = \tfrac{e}{2}\Delta V\left(1 - 2\tfrac{n+L_l}{L_s}\right) \quad \text{if } n \in \text{Sample}\ , \qquad (4)$$

where ΔV measures the bias applied to the contacts. A similar scheme was already used in computer simulations by Poppescu et al [12] on a different context.

As referred in Sect. 1, we will consider two setups having different initial configurations and perturbations. Namely:

- ***Partitioned Setup:*** the sample and leads are initially disconnected, $w_{L_l-1}^{L_l}(t) = w_{L_l+L_S-1}^{L_l+L_S}(t) = w\Theta(t)$, and the external bias is constant in time, $v_n^{\rm ext}(t) = V_n$. The initial state of the system for $t < 0$, is obtained by filling up eigenstates of the isolated leads and central sample, obtained from $\mathcal{H}(t < 0)$, according to local chemical potentials: $E_F \pm \Delta V/2$ for the left/right lead and E_F for the central sample;
- ***Non-Partitioned Setup:*** the sample and leads are always connected, $w_{L_l-1}^{L_l}(t) = w_{L_l+L_S-1}^{L_l+L_S}(t) = w$, and the bias is applied only for $t > 0$, $v_n^{\rm ext}(t) = \Theta(t)V_n$. The initial state of the connected system is obtained by filling up the eigenstates of the full system, obtained from $\mathcal{H}(t < 0)$, according to a global chemical potential E_F.

Finally, the charge current flowing from site n to site $n+1$, is described by the operator:

$$\mathcal{I}_n^{n+1} = \frac{ew}{i\hbar}\left(c_{n+1}^\dagger c_n - c_n^\dagger c_{n+1}\right). \qquad (5)$$

3 The Numerical Method and Results

In this section, we outline the method used to numerically calculate the $\mathcal{I}_n^{n+1}(t)$, after the external perturbation is suddenly connected. Let us suppose that $|\Psi(0)\rangle$ is some arbitrary occupied state. Then, its time-evolution is simply given by

$$|\Psi(t)\rangle = e^{-i\mathcal{H}(t>0)t}\,|\Psi(0)\rangle\,, \qquad (6)$$

where $\mathcal{H}(t > 0)$ is always time-independent and already includes the perturbation. Instead of calculating this exponential by diagonalization ($O\left(N^3\right)$ effort), and since $\mathcal{H}$ has a bounded spectrum, we may expand it as a Chebyshev series [15], yielding

$$|\Psi(t)\rangle = \sum_n^{\infty} \frac{2}{1+\delta_{n,0}}\,(-i)^n\, J_n(\lambda t) T_n(\tilde{\mathcal{H}})\,|\Psi(0)\rangle\,, \qquad (7)$$

where $\tilde{\mathcal{H}} = \lambda^{-1}\mathcal{H}\,(t > 0)$ is the rescaled Hamiltonian with spectrum inside $]-1, 1[$. Now, the most expensive part of the computation lies in calculating $|\Psi_n\rangle = T_n(\tilde{\mathcal{H}})\,|\Psi\,(0)\rangle$, up to some truncation order M. This may be done recursively, using the well-known properties of the Chebyshev Polynomials, $T_0(x) = 0$, $T_1(x) = x$ and $T_{n+1}(x) = xT_n(x) - T_{n-1}(x)$. Alternatively,

$$|\Psi_0\rangle = |\Psi\,(0)\rangle, \quad |\Psi_1\rangle = \tilde{\mathcal{H}}\,|\Psi\,(0)\rangle \quad \text{and} \quad |\Psi_{n+1}\rangle = \tilde{\mathcal{H}}\,|\Psi_n\rangle - |\Psi_{n-1}\rangle\,. \tag{8}$$

The series of Eq. 7 is seen to behave well upon truncation, as long as one keeps $M \gtrsim 4\lambda t$, which makes the calculation of $|\Psi\,(t)\rangle$ an $O\,(tN)$ process, optimized by a minimal choice of λ. Finally, to calculate the local electric current as a function of time, for all $t > 0$, we may use the previous method in the expectation value of $\mathcal{I}_s^{s+1}$. This yields

$$I_n^{n+1}(t) = \sum_{\text{occupied }\alpha} \langle\psi_\alpha|\, e^{i\mathcal{H}(t>0)t}\, \mathcal{I}_n^{n+1}\, e^{-i\mathcal{H}(t>0)t}\, |\psi_\alpha\rangle\,, \tag{9}$$

where $|\Psi_\alpha\rangle$ is an initially occupied eigenstate of $\mathcal{H}\,(t < 0)$. Furthermore, Eq. (9) may be recast in the more useful form

$$I_n^{n+1}(t) = \frac{2ew}{\hbar}\text{Im} \sum_{\text{occupied }\alpha} \langle\phi_n(-t)\,|\Psi_\alpha\,\rangle\,\langle\Psi_\alpha\,|\phi_{n+1}(-t)\,\rangle\,, \tag{10}$$

where $|\phi_n\rangle = c_n^\dagger\,|0\rangle$ is the state of an electron located at site n. The obvious advantage of the previous equation is that the numerical time-evolution need only to be performed twice.

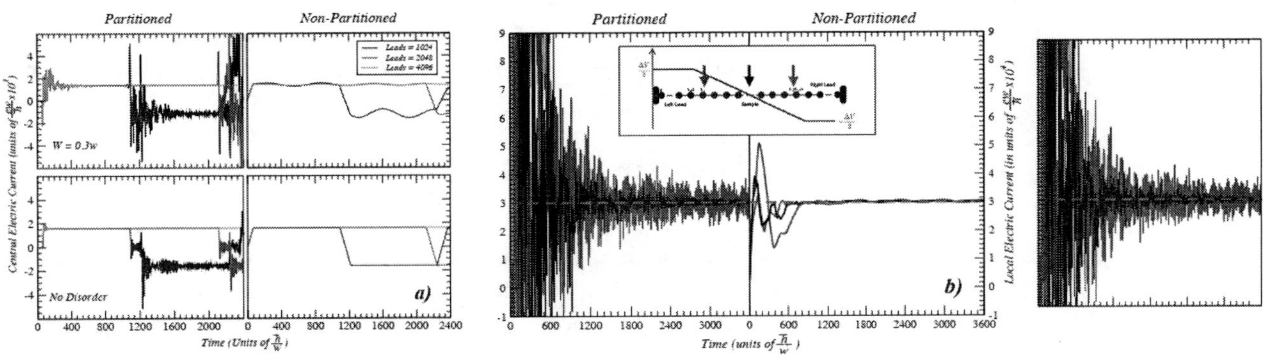

Figure 2. *a)* Plots of the time-dependent current across the central bond of a sample with $L_S = 256$ (with and without disorder), in both setups and increasing lead sizes. *b)* Plots of the current measured in the central and boundaries bonds (see legend) of a disordered sample ($W = 0.5w$) with $L_S = 512$ connected to leads with $L_l = 2^{16}$ sites. Both setups are also compared and the dashed magenta line are the values of the Landauer current obtained for this sample. In *a)* and *b)* the bias used was $\Delta V = 0.01w$.

In Fig. 2 a), we show the current across the sample's center as a function of time, for both initial setups, with or without disorder. In the *partitioned setup*, the results are similar to the ones obtained earlier for a multi-level system by Pal *et al* [9], where a quasisteady state Landauer current is found after a short transient. The Landauer current was obtained by writing the transmission function in Eq. (1) as $\mathcal{T}\,(E) = \gamma_L(E)\gamma_R(E)\left|\mathcal{G}^r_{0,L_s+1}\right|^2$, with $\gamma_{L/R}(E) = \sqrt{4w^2 - (E \mp e\Delta V/2)}$ and the retarded Green's function of the specific sample, in contact with semi-infinite leads, is evaluated using the *recursive green function algorithm* [8, 17].

The main novelty is that the same current is now obtained using the *non-partitioned* setup, which numerically shows that, for large but finite leads, the quasisteady value of the current is independent of the initial conditions. Hence it agrees with the proof given in [13] for the case of infinite leads.

However, this quasisteady state is followed by a current inversion (perfect in the ordered case) at a time t_r. The reflection of the propagating wavefunctions at E_F by the boundaries is conjectured to cause these inversions, seen in all plots of Fig. 2 a), which is confirmed by fact that $t_r \approx v_F^{-1} L_l$. Furthermore, since we do local measurements, we are also able to access the spacial profile of the current in the sample. In Fig. 2 b), we show that the quasisteady state current is indeed uniform, in the limit of very large leads and for both setups.
Nevertheless, we also report the existence of sample- and setup-specific oscillations of the quasisteady state current which are due to finite-size effects and disappear as L_l and t are increased. This is visible in the upper left panel of Fig. 2.
A definitive proof that a Landauer current is always established, for all biases, is provided by comparing the $I(V)$ curves of random samples as obtained from the quasisteady state current of our method, with the ones obtained from Eq. (1). The results are shown in Fig. 3, where one sees that all the previous examples were well beyond a linear transport regime.

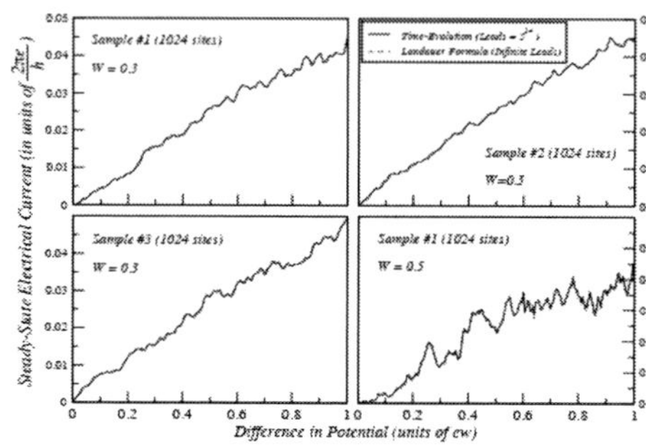

Figure 3. $I(\Delta V)$ curves of four disordered samples. The black curves were obtained from the quasisteady state regime of quantum dynamics in the non-partitioned setup with $L_l = 2^{14}$ sites. The red dashed curves were obtained from the Landauer formula [Eq. (1)]. $E_F = 0$ is assumed.

4 Summary and Conclusions

In this work, we studied the onset of quasisteady state transport regimes in large, but finite systems. Using a prototypical 1D tight-binding model, we have seen that the time-dependent current reaches a quasistationary value, even for finite leads. This quasisteady state regime, occurs at intermediate times, after transient dynamics have died out, but before the recurrent current inversions occur, due to the reflection of the electron wavefunctions at the boundaries.

As the size of the leads increases, so thus this recurrence time. During the quasisteady state regime, the current profile is nearly homogeneous as expected, and the value of the current in this quasisteady state regime coincides with the Landauer value, for large enough leads. Importantly, this quasisteady state regime is also independent on the initial condition of the system. We have thus extended previous results, valid only for infinite leads, on the emergence of a stationary Landauer current [3] and the independence of this state on the initial conditions [13] to systems with large, but finite leads.

Finally, the authors remark that these results and further developments on the time-dependent particle transport with "finite leads" have been published recently in Santos Pires *et al* [11].

Acknowledgements

While doing this work, J.P.S.P. was supported by the MAP-fis PhD grant PD/BD/142774/2018 of Fundação da Ciência e Tecnologia (FCT), Portugal. B.A. also acknowledges financing from FCT through Project Nº CEECIND/02936/2017. The authors jointly acknowledge financing from FCT and COMPETE 2020 program in FEDER component (European Union), through the projects POCI-01-0145-FEDER-028887 and UID/FIS/04650/2013.

References

[1] A. Ferreira and E. R. Mucciolo. *Phys. Rev. Lett.*, 115:106601, 2015.

[2] D. S. Fisher and P. A. Lee. *Phys. Rev. B*, 23(12):6851–6854, 1981.

[3] A.-P. Jauho, N. S. Wingreen, and Y. Meir. *Phys. Rev. B*, 50(8):5528–5544, 1994.

[4] S. M. João and J. M. Viana Parente Lopes. *Journal of Physics: Condensed Matter*, 32(12):125901, 2019.

[5] Simão M. João, Miša Anđelković, Lucian Covaci, Tatiana G. Rappoport, João M. V. P. Lopes, and Aires Ferreira. *Royal Society Open Science*, 7(2):191809, 2020.

[6] R. Kubo. *J. Phys. Soc. Japan*, 12(6):570–586, 1957.

[7] R. Landauer. *Phil. Mag.*, 21(172):863–867, 1970.

[8] A. MacKinnon. *Z. Phys. B: Cond. Matt.*, 59(4):385–390, 1985.

[9] P. P. Pal, S. Ramakrishna, and T. Seideman. *Jour. Chem. Phys.*, 148(14):144707, 2018.

[10] D. J. Passos, G. B. Ventura, J. M. Viana Parente Lopes, J. M. B. Lopes dos Santos, and N. M. R. Peres. 97(23):235446, 2018.

[11] J. P. Santos Pires, B. Amorim, and J. M. Viana Parente Lopes. Landauer transport as a quasisteady state on finite chains under unitary quantum dynamics. arXiv:1909.12402 (Accepted in *Phys. Rev. B*), 2020.

[12] B. S. Popescu and A. Croy. *Phys. Rev. B*, 95(23):235433, 2017.

[13] G. Stefanucci and C.-O. Almbladh. *Phys. Rev. B*, 69(19):195318, 2004.

[14] A. D. Stone and A. Szafer. *IBM Jour. Res. Develop.*, 32(3):384–413, 1988.

[15] H. Tal-Ezer and R. Kosloff. *Jour. Chem. Phys.*, 81(9):3967–3971, 1984.

[16] G. B. Ventura, D. J. Passos, J. M. B. Lopes dos Santos, J. M. Viana Parente Lopes, and N. M. R. Peres. *Phys. Rev. B*, 96:035431, 2017.

[17] Michael Wimmer. *Quantum transport in nanostructures: from computational concepts to spintronics in graphene and magnetic tunnel junctions.* Number 5 in Dissertationsreihe der Fakultät für Physik der Universität Regensburg. Univ.-Verl. Regensburg, 2009.

Probing the Global Delocalization Transition in the de Moura-Lyra Model with the Kernel Polynomial Method

N. A. Khan[,*], *J. P.* Santos Pires[,**], *J. M.* Viana Parente Lopes[,], and *J. M. B.* Lopes dos Santos[,]

[1]Centro de Física das Universidades do Minho e Porto Departamento de Física e Astronomia, Faculdade de Ciências, Universidade do Porto, 4169-007 Porto, Portugal

Abstract. In this paper, we report numerical calculations of the localization length in a non-interacting one-dimensional tight-binding model at zero temperature, holding a correlated disorder model with an algebraic power-spectrum (de Moura-Lyra model). Our calculations were based on a Kernel Polynomial implementation of the Thouless formula for the inverse localization length of a general nearest-neighbor 1D tight-binding model with open boundaries. Our results confirm the delocalization of all eigenstates in de Moura-Lyra model for $\alpha > 1$ and a localization length which diverges as $\xi \propto (1-\alpha)^{-1}$ for $\alpha \to 1^-$, at all energies in the weak disorder limit (as previously seen in [12]).

1 Introduction

The kernel polynomial method (KPM) [1] is now recognized as a valuable tool for calculating the physical properties of quantum systems and has been extensively studied in computational condensed-matter physics [2]. For our particular interest, this method has been successfully applied to compute the density of states [3], the inverse localization length [4] and even the spectral function of disordered one-dimensional systems, which can hold spatially-correlated disorder models [5].

It is well established that electronic states in the one-dimensional Anderson model are localized for all energies and any amount of disorder [6–8]. But for spatially correlated disorder, other possibilities arose. In 1998, de Moura and Lyra [10] analyzed an 1D tight-binding model with spatially correlated on-site disorder having an algebraic power-spectrum, $S(k) \propto k^{-\alpha}$, uncovering an unexpected insulator-to-metal transition as $\alpha = 2$ is crossed from below, accompanied by the appearance of a mobility edge which separates extended states (near the band center) from localized ones.

This model was revisited recently [12], having been shown to suffer from rather strong and tricky to handle finite-size effects over the entire parameter region $\alpha \in [0, \infty[$. For $\alpha > 1$, any finite section of the system becomes ordered in the thermodynamic limit, because the disorder's correlator for any two sites at distance r becomes a function of r/N, where N is the system size [11]. By devising a way to control finite size effects, and suitably take the thermodynamic limit, a delocalization transition of all eigenstates was proved to happen at $\alpha = 1$ in the thermodynamic limit [12]. This work measured the localization length using

*e-mail: nak@fc.up.pt
**e-mail: up201201453@fc.up.pt

the scaling of the typical mesoscopic conductance with system size, as calculated from the Landauer formalism.

In this paper, we use the KPM to determine the localization length, based on an old-result by Thouless, which is valid for 1D systems in the thermodynamic limit [6]. This work is a sequel of previous work by Santos Pires *et al* [12], which presents an independent verification of its results, and also demonstrates the usefulness of the KPM for studying localization properties on finite disordered lattices, even in the presence of strong finite-size effects. An excellent agreement is found between the two methods.

The following text is organized as follows. In Sec. 2, we present the tight-binding model of non-interacting fermions with onsite disorder, and review both Thouless's [6] and Izrailev and Krokhin's [13] results for the localization length of disordered 1D systems. In Sec. 3, we compute of localization length for the disordered systems of our interest using KPM implementation of Thouless's formula. We discuss our findings in Sec. 4 and summarize our conclusions in Sec. 5.

2 Disorder Models and Formulas for the Localization Length

A system of free fermions on an one-dimensional lattice with N sites has the following general tight-binding Hamiltonian

$$\hat{\mathcal{H}} = \sum_{i} \varepsilon_i \hat{c}_i^\dagger \hat{c}_i - \sum_{\langle ij \rangle} \left(t_{ij} \hat{c}_i^\dagger \hat{c}_j + h.c \right). \tag{1}$$

The essential parameters in the model are the hopping integrals t and the on-site values of the disordered potential, ε_i. For simplicity we will consider that only nearest-neighbor hoppings are relevant, i.e. $t_{ij} = t = 1$, and all energy scales will be measured in units of t. For the Anderson model, the on-site energies ε_i are independent random variables uniformly distributed in the interval $[-\frac{W}{2}, \frac{W}{2}]$, where W controls the strength of disorder. For the de Moura-Lyra model of correlated disorder, the randomness is long-range correlated with a power-spectrum $S(k) \sim k^{-\alpha}$ which decays algebraically with an exponent $\alpha > 0$. Hence, α is the parameter which loosely controls the roughness of potential landscapes in real-space. More precisely, the on-site energies are defined by [5, 10],

$$\varepsilon_i = \mathcal{A}_\alpha \sum_{k=1}^{N/2} \frac{1}{k^{\alpha/2}} \cos\left(\frac{2\pi k}{N} i + \phi_k\right). \tag{2}$$

where $\mathcal{A}_\alpha$ is the normalization constant that ensures a fixed variance of local disorder, and ϕ_k are $N/2$ independent random phases uniformly distributed in the interval $[0, 2\pi]$. In the limit $\alpha \to \infty$, this potential becomes a cosine potential with vanishing noise and a random overall phase. In this case, there are clearly eigenstates which are delocalized over the whole sample due to the effective absence of disorder. One also recovers Anderson's model of white noise potential (with a gaussian local distribution)in the limit $\alpha \to 0^+$.

As shown in Ref. [12], the localization of eigenstates in the de Moura-Lyra model mostly depends on the magnitude of the short-scale noise, rather than the power-law tails of the disorder correlator. The normalized single-bond discontinuity, defined as

$$D_N(\alpha, 1) := \frac{\left\langle (\varepsilon_n - \varepsilon_{n+1})^2 \right\rangle}{2\sigma_\varepsilon^2} = 1 - C_N(\alpha, 1), \tag{3}$$

measures the variance of the nearest-neighbor on-site energy differences for a correlated disorder realization with N values. In Eq. (3), $C_N(\alpha, 1)$ is the size-dependent two-point space correlator of the local disorder and, in the thermodynamic limit ($N \to \infty$), the normalized single-bond discontinuity for $\alpha < 1$ reduces to

$$D_\infty(\alpha, 1) = 1 - {}_1F_2\left(\frac{1-\alpha}{2}; \frac{1}{2}, \frac{3-\alpha}{2}; -\frac{\pi^2}{4}\right). \tag{4}$$

where ${}_1F_2(x)$ is an hypergeometric function.

The Thouless Formula:

In the Thouless's 1972 paper [6], an analytical expression for the characteristic decay length of the single-particle Green's function for a 1D model Hamiltonian with nearest-neighbor hoppings, open boundaries and holding an arbitrary on-site disorder. In the thermodynamic limit, the Thouless formula may be expressed in terms of the density of states of the disordered system —$\rho(\epsilon)$— as follows:

$$\frac{1}{\xi(E)} = \int \rho(\epsilon) \ln|E - \epsilon| \, d\epsilon. \tag{5}$$

This formula is rather general and does not involve any perturbative approach.

The Izrailev-Krokhin Formula:

Meanwhile, in the weakly disordered regime and in the thermodynamic limit, $\xi^{-1}(E)$ can also be calculated by a perturbative formula first derived by Izrailev and Krokhin [13], for a general model of correlated disorder. Their result, in the first-order on the local variance σ_ε, reads

$$\frac{1}{\xi(k)} \approx \frac{\sigma_\varepsilon^2}{8\sin^2 k}\left(1 + 2\sum_{r=1}^{\infty} C(r)\cos 2kr\right), \tag{6}$$

where $C_\alpha(r)$, is the normalized two-point correlator of the random potential and $E = 2\cos k$. Applying this to our two models of interest one gets: *(a)* $\xi \approx 105.2/W^2$,for the Anderson model, at the band center; [8] And *(b)*,

$$\xi(E) = \frac{8}{(1-\alpha)\sigma_\varepsilon^2}(1 - \frac{E^2}{4})\left[\frac{2}{\pi}\arccos(\frac{E}{2})\right]^\alpha. \tag{7}$$

for the de Moura-Lyra model [12], at an energy E. The expression of Eq. (7), can be used to calculate the localization length of the de Moura-Lyra model for any values of the energy and of the correlation controlling parameter α, in limit $N \to \infty$.

3 Kernel Polynomial Method implementation of the Thouless Formula

The kernel polynomial method [1] is a polynomial expansion-based technique for calculating spectral quantities with a computational effort that scales linearly with system size. The Chebyshev polynomials are a convenient choice for expanding the desired quantities, due to the convergence properties of their summations and its relation to Fourier transforms. The first kind Chebyshev polynomials, $T_m(x)$, are defined as

$$T_m(x) = \cos(m \arccos(x)), \quad m \in \mathbb{N}. \tag{8}$$

These polynomials obey the following recursion relation

$$T_m(x) = 2xT_{m-1}(x) - T_{m-2}(x), \qquad m > 1. \tag{9}$$

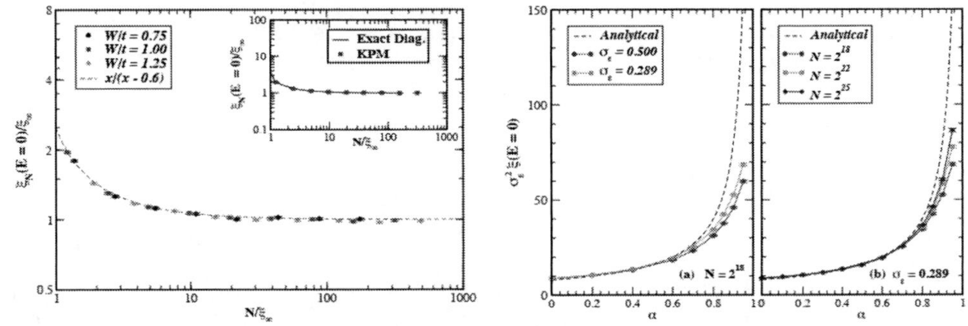

Figure 1. Left panel: Numerical study of the convergence of the localization length to its thermodynamic limit value, i.e. plot of $\xi_N(E=0)/\xi_\infty$ as a function of N/ξ_∞ for the 1D Anderson model with open boundaries. The data were computed using the KPM with $M = 2048$ Chebyshev moments and have an estimated error of 1%. The parameter ξ_∞ is the analytically calculated localization length in the perturbative regime for $N \to \infty$ [8]. In the inset, we confirm the convergence of the KPM estimates of the localization length by comparing with the results calculated by an exact diagonalization of the disordered Hamiltonian for $W/t = 1$.
Right panel: Estimates of the normalized localization length $\sigma_\varepsilon^2\xi(E=0)$, as a function of α for the de Moura-Lyra model with OBC using the KPM. The computations were carried out for a fixed *(a)* system of size of$N = 2^{18}$sites and various local variances,σ_ε, and *(b)* $\sigma_\varepsilon = 0.289$ (i.e. $W = t$) and for various system sizes. Once again, $M = 2048$ Chebyshev moments were used to obtain an 1% estimated error of convergence. The dashed black curve stands for the Izrailev-Krokhin result (Eq. (7)) and, in *(b)*, the numerical curves seem to be converging towards this one as N is increased. This reflects the strong finite-size effects which are known to exist in this problem.

which starts with $T_0(x) = 1$ and $T_1(x) = x$. Additionally, they also satisfy the orthogonality relation

$$\int_{-1}^{1} T_n(x)T_m(x)(1-x^2)^{-1/2}dx = \frac{\pi}{2}\delta_{n,m}(\delta_{n,0}+1). \tag{10}$$

Since Thouless' result [6] expresses the inverse localization length in terms of a density of states, KPM estimates of the localization length may be obtained from the KPM approximated density of states [1, 3] as: [4]

$$\frac{1}{\xi(E)} = -\ln 2 - 2\sum_{m=1}^{M-1}\frac{\mu_m g_m}{m}T_m(E). \tag{11}$$

While this allows the computation of $\xi^{-1}(E)$ for any band energy E, we will focus on the band center without any loss of generality. The KPM moments in Eq. 11, μ_m, are defined as

$$\mu_m \;=\; \int_{-1}^{1} T_m(\epsilon)\rho(\epsilon)\,d\epsilon = \tfrac{1}{N}\mathrm{Tr}[T_m(\hat{\mathcal{H}})]. \tag{12}$$

The expression of Eq. (11) is the truncated series with M terms. This abrupt truncation introduces Gibbs oscillations—signaling non-uniform convergence. This numerical artifact can be filtered out of the final result by employing an optimized damping factor, such as the so-called **Jackson Kernel** g_m defined as [1]:

$$g_m = \frac{(M-m+1)\cos(\frac{m\pi}{M+1})}{M+1} + \frac{\sin(\frac{m\pi}{M+1})\cot(\frac{\pi}{M+1})}{M+1}. \tag{13}$$

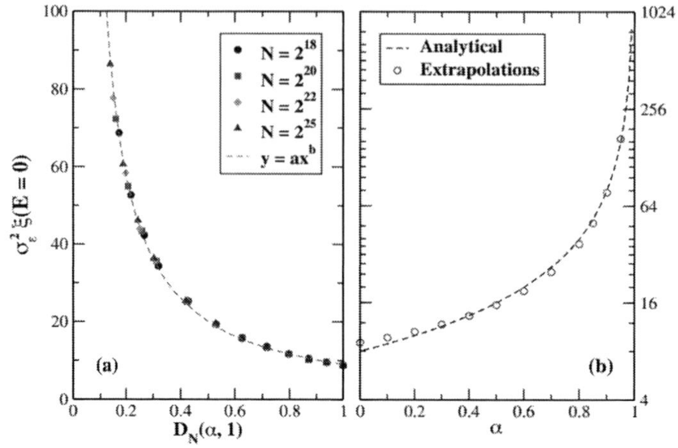

Figure 2. *(a)* Plot of the normalized localization length, $\sigma_\varepsilon^2\xi(E = 0)$, as a function of the parameter $D_N(\alpha, 1)$, for the de Moura-Lyra model with OBC at zero temperature. The numerical computations were carried out for a disorder variance $\sigma_\varepsilon = 0.289$ using $M = 2048$ Chebyshev moments and an estimated error of 1%. The data can be collapsed into a single curve $y = ax^b$(dashed magenta line). *(b)* Comparison between the thermodynamic limit values of $\sigma_\varepsilon^2\xi(E = 0)$ predicted by the dashed curve in (b), with the analytical expression of Eq. (7).

4 Results and Discussion

In order to validate our KPM estimates of ξ^{-1}, we first addressed the numerical convergence of the localization length for the well-known uncorrelated Anderson model. In the left panel of Fig. 1, we illustrate the convergence of our finite lattice results towards the perturbative thermodynamic limit result ($\xi = 105.2/W^2$) as the size of the chain is increased, for different disorder strengths. All the calculations were done with an estimated error of 1% as determined by the standard-deviation of the sample-specific fluctuations in the localization length. In addition, we found the numerical data of Fig. 1 to be well fitted to the scaling function,

$$\frac{\xi(E=0)}{\xi_\infty} = \frac{x}{x - x_0}, \qquad x_0 \sim 0.6. \tag{14}$$

where $x = N/\xi_\infty$.

The delocalization behavior of the eigenstates in the de Moura-Lyra potential is illustrated in Fig. 1 (right panel). The rescaled localization length shows large deviation from Izrailev's result for $\sigma_\varepsilon = 0.5$ (non-perturbative regime). However, it starts to converge with decreasing σ_ε for a fixed-system size as signaled by the approaching curves in *(a)*. Nevertheless, there are strong system finite-size effects [12], clearly visible in (b), which prevent the observation of a fully converged curve for sizes as large as $N = 2^{25}$sites. The convergence is then very slow near the transition point $\alpha \sim 1$.

The Fig. 2*(a)* illustrates the scaled localization length of the model as a function of the parameter $D_N(\alpha, 1)$, instead of α, in the perturbative regime ($\sigma_\varepsilon = 0.289$). We now find a perfect collapse of all the data into a curve of the form $y = ax^b$(dashed magenta line). With this collapse, we are able to understand the operation of these finite-size effects — by increasing the system size, all the data points ($\sigma_\varepsilon^2\xi(E = 0)$) slide upwards over the dashed curve Finally, we can read the thermodynamic limit value of $\sigma_\varepsilon^2\xi(E = 0)$ for a given α, as the ordinate of the dashed curve at $x = D_\infty(\alpha, 1)$ (see Eq. (4)). A comparison with the analytical result (Eq. (7)) is depicted in Fig. 2*(b)*. One can clearly see that the extrapolated data shows an excellent agreement with the analytical result, thus confirming the delocalization transition at $\alpha = 1$ found earlier by the authors.

5 Conclusions

We reported numerical calculations of the localization length in one-dimensional disordered tight-binding models with uncorrelated (Anderson model) and also spatially correlated (de Moura-Lyra model) on-site disorder. The calculations were done by applying the very efficient Kernel Polynomial Method to implement the Thouless formula for the localization length of a general one-dimensional tight-binding model with open boundaries.

For the Anderson model, we found an excellent agreement of the KPM-estimated localization length with the perturbative analytical results, confirming the validity of our procedure. We also studied the dependence of the measured ξ with the chain's size, thus checking its convergence to the well-known value in the thermodynamic limit.

For the trickier de Moura-Lyra model, we showed that our KPM implementation of the Thouless formula fully confirms results obtained previously with Landauer formalism.Namely, we were able to reproduce the divergent behavior of the localization length , $\xi \propto (1-\alpha)^{-1}$, in the vicinity of delocalization transition at "$\alpha = 1$", in the perturbative regime. This observation was done by a finite-size extrapolation similar to the one used by Santos Pires *et al* [12].

Acknowledgments:

For this work, N.A. Khan was supported by the INTERWEAVE project, Erasmus Mundus Action 2 Strand 1 Lot 11, EACEA/42/11 Grant Agreement 2013-2538/001-001 EM Action 2 Partnership Asia-Europe and J. P. Santos Pires was supported by the MAP-fis PhD grant PD/BD/142774/2018 of Fundação da Ciência e Tecnologia. The authors also acknowledge financing from Fundação da Ciência e Tecnologia and COMPETE 2020 program in FEDER component (European Union), through the projects POCI-01-0145- FEDER-028887 and UID/FIS/04650/2013.

References

[1] A. Weiße, G. Wellein, A. Alvermann, and H. Fehske, Rev. Mod. Phys., 78:275 (2006).
[2] F. Lacopi, J. J. Boeckl, and C. Jagadish, *2D Material* (Academic Press, 2016)
[3] R. N. Silver, H. Roeder, A. F. Voter, and J. D. Kress, J. Comput. Phys. 124:1 (1996).
[4] N. Hatano and J. Feinberg. Phys. Rev. E, 94:063305 (2016).
[5] N. A. Khan, J. M. Viana Parente Lopes, J. P. Santos Pires, and J. M. B. Lopes dos Santos. J. Phys.: Condens. Matter, 31:175501 (2019).
[6] D. J. Thouless. J. Phys. C: Solid State Phys., 5:77 (1972).
[7] D. J. Thouless. *Ill-Condensed Matter*, Les Houches Session XXXI. (North-Holland, New York, 1979).
[8] F. M. Izrailev, A. A. Krokhin, and N. M. Makarov. Phys. Rep., 512:125 (2012).
[9] P. A. Lee and T. V. Ramakrishnan. Rev. Mod. Phys., 57:287 (1985).
[10] F. A. B. F. de Moura and M. L. Lyra. Phys. Rev. Lett., 81:3735 (1998)
[11] G. M. Petersen and N. Sandler. Phys. Rev. B, 87:195443 (2013).
[12] J. P. Santos Pires, N. A. Khan, J. M. Viana Parente Lopes, and J. M. B. Lopes dos Santos. Phys. Rev. B, 99:205148 (2019).
[13] F. M. Izrailev and A. A. Krokhin. Phys. Rev. Lett., 82:4062 (1999).
[14] D. Zwillinger. *Table of Integrals, Series, and Products*. (Academic Press, 2014).

AUTHOR INDEX

EDP Sciences – Web of Conferences
17, Avenue du Hoggar
Parc d'Activité de Courtabouf
BP 112
F-91944 Les Ulis Cedex A
France

ISBN 978-1-7138-1016-2

CURRAN ASSOCIATES INC.
proceedings.com